MARINE FISHES
AND
INVERTEBRATES
IN YOUR OWN HOME

By Dr. Cliff W. Emmens

t.f.h.

PHOTO CREDITS

The author is most appreciative of the many photographers who have helped illustrate this book. Not every photo or drawing submitted was used, so thanks goes to those people as well.

The following persons are deserving of special mention:
Dr. Herbert R. Axelrod, Bill Chung, Dr. Gerald Allen, M. Goto, Herr Frickhinger, Prof. Dr. Reichenbach-Klinke, Dieter Untergasser, Rodney Jonklaas, Helmut Debelius, Allan Power, Jim Greenwald, Walter Deas, Dr. Denis Terver, John Burleson, Firma Dupla, Aqua Module, Robert Straughan, Scott Johnson, C.O. Masters, Dr. Fujio Yasuda, Dr. Karl Knaack, Douglas Faulkner, Gerhard Marcuse, H. Hansen, Hiroshi Takeuchi by courtesy of Midori Shobo Fish Magazine, Dr. Harry Huizinga, Dr. J.E. Harris, Dr. D.H. Lewis, D. McGregor, Dr. Elkan, Dr. Mark Dulin, A. Norman, Ken Lucas, Dr. Patrick Colin, Jan Carlin, William M. Stephens, Walter Starck II, U. Erich Friese, Peter T. Jam, Roger Steene, Ron Thresher, Georg Smit, Bruce Carlson and P. Frankbonner. The listing is done in random order and there is no significance to the appearance of any name's listing before any other name. Actually, the names appear in the order in which I accepted the photos.

I am almost certain to have forgotten to thank some photographer, so please advise me if I left out a reference inadvertently.

I wish to thank John Quinn of *Tropical Fish Hobbyist Magazine* for the drawings he made. All drawings are basically from John. I also wish to thank Jerry Walls and Dr. Herbert R. Axelrod for the layout and captions.

C.W. Emmens
Sydney, Australia

1995 Edition

Distributed in the UNITED STATES to the Pet Trade by T.F.H. Publications, Inc., One T.F.H. Plaza, Neptune City, NJ 07753; distributed in the UNITED STATES to the Bookstore and Library Trade by National Book Network, Inc. 4720 Boston Way, Lanham MD 20706; in CANADA to the Pet Trade by H & L Pet Supplies Inc., 27 Kingston Crescent, Kitchener, Ontario N2B 2T6; Rolf C. Hagen Ltd., 3225 Sartelon Street, Montreal 382 Quebec; in CANADA to the Book Trade by Vanwell Publishing Ltd., 1 Northrup Crescent, St. Catharines, Ontario L2M 6P5 ; in ENGLAND by T.F.H. Publications, PO Box 15, Waterlooville PO7 6BQ; in AUSTRALIA AND THE SOUTH PACIFIC by T.F.H. (Australia), Pty. Ltd., Box 149, Brookvale 2100 N.S.W., Australia; in NEW ZEALAND by Brooklands Aquarium Ltd. 5 McGiven Drive, New Plymouth, RD1 New Zealand; in Japan by T.F.H. Publications, Japan—Jiro Tsuda, 10-12-3 Ohjidai, Sakura, Chiba 285, Japan; in SOUTH AFRICA by Lopis (Pty) Ltd., P.O. Box 39127, Booysens, 2016, Johannesburg, South Africa. Published by T.F.H. Publications, Inc.
MANUFACTURED IN THE UNITED STATES OF AMERICA
BY T.F.H. PUBLICATIONS, INC.

Table of Contents

PREFACE 5

CHAPTER 1
MARINE ENVIRONMENT 7

CHAPTER 2
SOME MEASUREMENTS 19

CHAPTER 3
CONTROLLING AQUARIUM CONDITIONS 25

CHAPTER 4
BIOLOGICAL FILTRATION 41

CHAPTER 5
SETTING UP YOUR MARINE AQUARIUM 53

CHAPTER 6
BUYING AND HANDLING FISHES 63

CHAPTER 7
FEEDING MARINE FISHES 76

CHAPTER 8
DISEASES AND PARASITES 97

CHAPTER 9
THE NATURAL SYSTEM 133

CHAPTER 10
REPRODUCTION IN MARINE FISHES 163

BIBLIOGRAPHY 190

INDEX 191

ABOUT THE AUTHOR

Dr. Cliff Emmens is both a zoologist and physiologist by profession and an aquarist as a lifelong hobby. At one time he had as many as 70 tanks of freshwater and marine fishes. He set up the first veterinary physiology department in Australia, which became a blueprint for subsequent departments in other universities, and was its chairman from 1948 to 1978, when he retired as an Emeritus Professor with the degrees of Ph.D., D.Sc., and D. V. Sc.

In the course of a long career, Cliff Emmens has been President of the Australasian Region of the International Biometric Society, the Endocrine Society of Australia, the Australian Society for Reproductive Biology, the Sydney Association of University Teachers, and the Second Asia and Oceania Congress of Endocrinology. In addition he has been on the councils of various other societies, a member of many committees, and consultant to various firms such as Mead Johnson and Syntex.

Emmens is the author of several biological and statistical texts and author or coauthor of ten books on aquarium keeping, several with Dr. Herbert R. Axelrod. His contributions to the scientific literature in the form of original research papers and chapters in textbooks exceed two hundred, while his classified papers on the techniques of bombing and their results add a further twenty or so. During World War II he travelled extensively in the North African and European theatres with the honorary rank of Wing Commander in the Royal Air Force. His awards include the Oliver Bird Medal & Prize (U.K.), the Italian Istituto Spallanzani Medal, 5 campaign and other war medals, election to the Australian Academy of Science, and an honorary Fellowship of the Australian College of Veterinary Scientists.

Dr. Cliff W. Emmens

Marine aquarium keeping is one of the most rewarding of hobbies. You can make it as simple or as complicated as you wish. At its simplest it is little more difficult than the freshwater aquarium hobby. Modern tanks and equipment take most of the hassle out of it, and advice is now freely available from your local pet shop on the various ways of setting up and running a successful marine aquarium. The easiest and most trouble-free ways are discussed in this book.

Beyond merely keeping marine fishes there are many other possibilities, most of them only in the early stages of development. There is the question of higher algae and their cultivation, so far only touched upon. Then comes the breeding of both fishes and invertebrates and the problem of raising the young. Many years will no doubt pass before we can breed marine fishes as easily as we do freshwater ones, but a very promising start has been made and there is even commercial production of a few species. There is the great range of invertebrates, many of which are

kept successfully in home aquaria, but with many others there are unsolved problems. Most invertebrates go through complicated life histories, but some of them can be bred if the necessary trouble is taken—indeed, we have to breed a few of them to feed our larval fishes, although other diets may eventually be substituted. Invertebrates such as some anemones and free-living copepods breed readily without our help, and others that cause disease also do so despite our efforts to suppress them.

A neglected section of the hobby is keeping temperate zone aquaria. Most of us live in the temperate zones, but few keep aquaria with local fishes and invertebrates, even though many of them are attractive and some are even as attractive as tropical species. Exactly the same techniques are needed as for tropicals. Lower heating costs may repay installation in winter, and the only problem that may arise in some areas is overheating in summer. This may be overcome by adapting the works

of a discarded refrigerator or by evaporative cooling, but so many people have air-conditioning that they will find it no problem. The great advantages of a "cold" aquarium are the low cost of the specimens (if you live near or can visit the coast) and the fun of collecting them. Such an aquarium needs less attention than a tropical one as the metabolism of the inhabitants is

It is actually possible to take a piece of living coral rock, bring it into your home aquarium, and watch it grow. By adding a few compatible fishes, you can have a mini-reef aquarium in your own living room. Drawn by John Quinn.

slower and the oxygen content of the water higher, so the margin of safety is greater. You won't find special reference to it in this book because nothing more needs to be said. A temperate "natural system" aquarium is especially recommended.

I shall not start this work like the usual books on aquaria, with a description of the tank and its

Beginners should start their mini-reef or marine aquarium with hardy fishes and easy-to-keep invertebrates. This book will help you through the maze of questions and problems that might arise as you set out on your new hobby.

equipment, then chapters on maintenance, feeding, and so on. Instead, I shall discuss the ocean, its constituents, and substitutes for it, and the measurement of important factors in the aquarium and how to control them. With this as a background it is easier to see why particular types of aquaria and ways of setting them up are recommended.

Finally, a word to beginners. Start by keeping some tough fishes like damsels (including anemonefishes), wrasses, small triggers, gobies, and blennies and leave the harder-to-keep varieties, particularly chaetodons and angels, until later. Then try

some of the easier invertebrates like anemones, starfishes, crabs, shrimps, and tubeworms, preferably in a tank of their own. I suggest trying a smallish tank with a large anemone and a pair of anemonefishes, and finally graduating to a mixed display of fishes, invertebrates, and algae that will be the envy of your friends and a demonstration that you have really arrived as a marine aquarist. When you feel at home with marine aquaria, only then is it the time to think of trying the natural system or miniature reef and making attempts at breeding.

C. W. Emmens

The ocean is usually described by writers for aquarists as having a remarkably constant composition, being subject to slow and small temperature changes, and in general offering a relatively unchanging environment in which its denizens live. This is perhaps true of the great depths (at any rate away from volcanic trenches) and of the mid-ocean at medium depths, but it is not true of the surface and the shores of the ocean. Pelagic fishes like the herrings do in fact live in relatively stable conditions—they swim at depths immune to surface squalls and rain, they do not frequent tidal areas, and they enjoy plenty of oxygen and a fairly even temperature as long as they do not penetrate the thermocline (a shifting zone usually 50-100 feet from the surface where the temperature and density change suddenly as you descend). Such fishes are not easy to keep in captivity, exactly because of their being accustomed to a stable environment.

None of this is true of the regions in which most of the common aquarium fishes live. They are collected from tidal zones and from shallow reef areas, very rarely from any great depths where conditions are stable. In consequence, they are sometimes subjected to abrupt changes in water temperature and density, as when it rains heavily or when tide pools are flooded out by the incoming tide, after having been heated by the sun for several hours and then cooled by water at a lower temperature. They do have good oxygenation and plenty of water movement most of the time, and given these they are happy in aquarium conditions that do not suit the herring. They are often accustomed to a degree of

pollution and turbidity that pelagic fishes avoid, and so can withstand tougher conditions than the latter. The same is true of many aquarium invertebrates, but there is of course a limit to what any organism can stand.

The Constituents of Seawater

To understand the requirements of a successful marine aquarium it is essential to know something about the make-up of the oceans and which of the many substances present are needed by their inhabitants. Ocean water is a very complex substance containing nearly all of the natural elements, some in only minute quantities. However, when we consider the vastness of the world's oceans, there are many tons of any one of them present in total. (Hence the interest in extracting gold from sea water, although it is at a concentration of only about one part in 250 billion.) The makeup of this complex mixture is usually divided into the major constituents (dissolved salts), the minor constituents or trace elements (also salts), dissolved gases, organic substances in solution, and organic and inorganic particles in suspension.

Dissolved Salts There are ten major constituents that vary very little in concentration in undiluted or unpolluted ocean water and account for over 99% of the dissolved material. These typically add up to a 3.5% (approx.) solution of salts (natural extremes 3.3–3.8%) containing the metals sodium (1.08%), magnesium (0.13%), calcium (0.041%), potassium (0.039%), and strontium (0.001%), together with chloride (1.94%), sulphate (0.27%), bromide (0.007%), and borate (0.003%) bases and a mix of carbonates, bicarbonates, and free carbon dioxide (about 0.003%). In certain enclosed areas, like the Red Sea, the percentage of salts may go as high as 4.0%, and in inshore areas it may fall to 3.0% or even less.

The minor constituents in their dozens only add a total of 0.0005% to the above, while dissolved oxygen gas is typically

If two fish are antagonistic to each other, such as a male and female of the same species during a pre-spawning ritual, it is possible to separate them with a tank divider. Your local aquarium petshop will have dividers for tanks and many other gadgets necessary for your success.

Pelagic fishes that are almost never found on the reef are very difficult to maintain in the home aquarium because they are not accustomed to rapid changes in temperature, water condition, and pollution. Reef fishes are, by nature, acclimated to such changes. These *Carangoides emburyi* were photographed by Allan Power.

at a concentration of 0.0008%, nitrogen at 0.0013%, and dissolved organic matter between 0.0001% and 0.0025%, depending on locality. Particles in suspension are naturally very variable in amount, around 0.0001% in clear ocean water but much higher in tidal zones.

It is impossible to say that there is so much sodium chloride, so much magnesium sulphate, etc. in sea water as the behavior of its contents shows that 80% are dissociated, which means that it is mostly a mixture of positively charged metallic ions (Na^+ (sodium), Mg^{++} (magnesium), etc.) and negatively charged basic ions (Cl^- (chloride), SO_4^{--} (sulphate), etc.) At any one time about 20% of the sodium ions will be combined with a base forming NaCl (sodium chloride), Na_2SO_4 (sodium sulphate), and so on in proportion to the numbers of Cl^-, SO_4^{--}, etc., ions available. The same is true of the other metallic ions. What has to

be done in making up a synthetic copy of sea water is simply to supply the needed amounts of sodium, magnesium, chloride, sulphate, etc., to give the same concentrations as in the natural article. Hence different formulae for making up a synthetic mix are quite usual; one may supply all the sodium as sodium chloride and all the magnesium as magnesium sulphate, while another may not.

As an example of a synthetic substitute for sea water, the oft-quoted simple mix of Lyman & Fleming (1940, *J. Marine Res, 3:134*) contained:

Sodium chloride (NaCl) = 23.48g
Magnesium chloride ($MgCl_2 \cdot 6H_2O$) = 4.98g
Sodium sulphate ($Na_2SO_4 \cdot 10H_2O$) = 3.92g
Calcium chloride ($CaCl_2$) = 1.10g
Potassium chloride (KCl) = 0.664g
Sodium bicarbonate ($NaHCO_3$) = 0.192g
Potassium bromide (KBr) = 0.096g
Boric acid (H_3BO_3) = 0.026g
Strontium chloride ($SrCl_2$) = 0.024g
Sodium fluoride (NaF) = 0.003g
All weights are grams per liter of water

Except for the sodium fluoride, often omitted in fluoridated areas, the mix contains only the major constituents and will

A marine aquarium is a complex chemical reaction that continually changes as the water evaporates, the fish excrete their waste, and you place food or additional chemical stress on the aquatic environment in the tank.

support fishes and some invertebrates fairly well, but for best results should have 10% natural sea water added to it.

Pet shops sell pre-mixed sea salts with 70 trace elements. These are, of course, better than the mix mentioned above.

By no means have all of the minor constituents been shown to be essential to marine life. It seems likely that most if not all fishes can survive in a mixture as above, getting all their other needs from the diet, just as we do, but the matter seems not to have been adequately investigated. What is certain is that many invertebrates and all seaweeds need some of the minor constituents. Those known to be essential to all algae are nitrogen, phosphorus, iron, copper, manganese, zinc, and molybdenum, while many others are needed by one or another species. The majority of algae also need vitamins B_{12} and

Keeping sharks in the home aquarium is becoming more and more popular. Most sharks are rather good aquarium inhabitants if they are properly fed and their tank is large enough. The juveniles of large sharks, like these *Carcarhinus melanopterus*, can be dwarfed simply by feeding them as little as possible when they are still small. Photo by S. Johnson.

thiamine. In addition to the list for algae, iodine, cobalt, vanadium, and arsenic are known to be needed by various invertebrates, some of which concentrate one or another trace element to a remarkable degree. Being present only in trace amounts, these substances are rapidly depleted in the aquarium, and for good algal growth it may be necessary to replace them at intervals. They are available with instructions for use from your local aquarium supplier.

Other Constituents Dissolved organic (carbon-containing) substances derived from living matter usually amount to around 5-6 parts per million (ppm) in northern seas, where most measurements have been taken, or quoted, and consist of proteins, polypeptides, and amino acids—the last two are products of protein breakdown or are protein building blocks,

according to how you look at it—plus traces of thiamine, biotin, and vitamin B_{12}. Particles in suspension in clear coastal waters off Plymouth, in the U.K., were found by Harvey to occur at 0.4 to 2 ppm, and from the Atlantic hundreds of miles from land, 0.2 to 1 ppm. About half of this particulate matter was inorganic iron-containing clay-like material. Turbid inshore waters also contain organic and inorganic phosphates in the particles. Such particles are not unimportant, as they provide a basis for the growth of bacteria that influence the organic constituents.

Quite apart from dilution by rainwater runoff, direct or in the form of estuaries, inshore water may and far too often does contain man-made pollutants. The worst of these are sewage, industrial wastes, agricultural sprays and fertilizers, and silt, which may clog marine life and kill off corals.

Many of these pollutants are toxic; others, like fertilizers, may cause algal "blooms" that can be toxic in turn to other marine life. The so-called red tides are an example of the latter and can poison fishes. Hence it is hazardous to collect inshore sea water for the aquarium, and it may be much safer to use a synthetic mix.

Plankton
The living contents of sea water vary enormously according to locality and time of day, but any sample from the surface of the oceans or inshore waters will contain a great number of planktonic organisms—the phytoplankton, consisting of floating, mainly one-celled plants and the spores of larger species; and the zooplankton, minute floating animals and the young stages of larger species. The zooplankton in particular tends

to rise to the surface at night and sink to a lower zone during the day. Bacteria are abundant inshore, but are quite sparse in the upper layers of the open ocean, dropping to as low as 10 per ml. Many of the larval stages of fishes, crustaceans (crabs, shrimps, etc.), coelenterates (anemones and corals, etc.), echinoderms (starfishes, sea urchins, etc.), and other invertebrates live in the plankton, feeding on the phytoplankton, other types of zooplankton, and each other. They are in turn eaten by larger organisms, particularly crustaceans, which in their turn feed many species of fishes.

This food chain is a factor in which the aquarium differs from the sea. The typical sea water aquarium does not have plankton, although it may have abundant bacteria. Bursts of larvae from the inhabitants of an invertebrate tank may occur, particularly in the so-called natural system tank, but they don't provide the variety and abundance of natural plankton although they may contribute briefly to the fishes' diet. The existence of normal plankton depends on contributions from a vast number of marine life forms that cannot be maintained in any but very large volumes of water and a variety of habitats. The aquarium is very definitely not the sea in miniature.

Stored Sea Water

As soon as sea water is collected it begins to change.

These fish (Stonogobiops sp.) are very rare and are hardly ever seen for sale in the usual aquarium shop. But a lucky aquarist who might be diving or snorkeling could chase them into a hole and capture them. Almost all reef fishes will do well in the aquarium unless they are too large or another fish attacks them.

Plankton dies, releasing organic material into the water, and bacteria proliferate. Just how fast they proliferate depends on the size of the container, the temperature, nutrients available, oxygen available, and various other factors not fully investigated. Algae may also proliferate if the water is well lit, but this is usually avoided by storage in the dark. The larger the container, the lower the peak bacterial content, as much of the bacterial growth depends on the availability of surfaces over which they multiply, so that the lower the surface/volume ratio of a container, surface meaning the inner walls and floor, the lower the bacterial count. The higher the temperature, the more rapid the growth of bacteria and depletion of oxygen in unaerated vessels. The peak usually occurs between 2 and 10 days from collection and involves a rise from a few dozen or a few hundred bacteria per ml to many thousands in large vessels to perhaps half a million or more in smaller vessels of only a few gallons.

In unaerated dark storage the dead plankton and some of the bacteria gradually settle to the bottom and the bacterial content of the water drops to a lower level once more, but rarely as low as it was to begin with. Counts of from 10–100,000 per ml have been reported. There will be no plankton and no protozoa that can cause disease; even bacterial causes of disease will mostly have died down, suppressed by the other bacterial growth and the absence of the tissues they usually infect. After two or three weeks the sea water is ready for use in the aquarium and can be carefully siphoned off, avoiding the slush at the bottom. Some collectors filter the water before storage, which helps to hasten things, but it would still be advisable to store even filtered water for at least two weeks.

From all of this it will be clear that there are two ways to use new sea water—immediately, but with a risk of disease-causing organisms being introduced to the tank, or after storage in the dark as just described. It is a temptation to use the water right away, and in many circumstances it is fairly safe to do so as the concentration of disease-causing organisms or parasites is very low in most waters, but it is safer to store it. Storage not only removes plankton but also gives some types of pollutants a chance to biodegrade. However, it also gives a chance for other types to encourage a rise in ammonia and nitrites that may poison the water for months, so this is no excuse for risking the collection of even mildly polluted water.

Collecting Sea Water

Collect only from the open ocean if at all possible as inshore waters tend to be polluted, as we have seen. If you have no choice and must collect from a harbor or tidal area, do so away from human activities as far as is possible and collect on a rising tide. This will give you as pure a sample as you are likely to get. If you collect after high tide, you will be sampling the run-off from the harborside or shore, which is likely to be highly polluted. Collect from an area where green algae, anemones, and general sea life are abundant and healthy looking—if these living forms can flourish the water is likely to be safe. Even so, they may be especially tough survivors and it is still much better to get water

In the "old days," Robert Straughan, one of the pioneers in marine home aquarium keeping, was very proud of this aquarium. However, he preached that all coral placed into the aquarium, including living rocks, should be boiled, bleached, and have everything that once lived removed before it is placed in the aquarium. On the facing page is a modern mini-reef aquarium.

Most petshops that sell and service marine aquariums are able to supply you with living rock. This will form the background and sides of your mini-reef aquarium. Additional specimens of algae, anemones, clams, etc., can be added to the reef, but the rocks will start to grow (they actually contain the eggs and immature specimens of many marine invertebrates) after a few weeks and provide some interesting living material. Very few fishes should be added until the rocks have developed as some fishes feed on many invertebrates...and many invertebrates (starfish, snails, shrimp, crabs, etc.) feed voraciously on other invertebrates.

from out to sea. At sea, make sure that there are no oil slicks, discharged raw sewage, or other signs of trouble; even the open ocean is not always safe nowadays.

Collect the water in inert plastic containers or other containers lined with strong plastic bags of guaranteed safe composition. Water is heavy so containers of about 5 gallons capacity are convenient, the

plastic jerrican being ideal, as is a plastic garbage can of not too large a size. The latter can be stored as is, since you can siphon easily from it without disturbing the silt at the bottom. The water in a jerry can has to be poured into another container for settling out, either immediately or eventually. Once home, see that the storage vessel is tightly covered to avoid its collecting airborne contaminants, and store

it in the dark or in a dark vessel for at least two, preferably three, weeks, preferably in a warm atmosphere to hasten bacterial growth and decline. Make a note on the vessel or elsewhere of the date of collection, otherwise— "let me see, when did we go out to collect this water?". When the water is ready for use, siphon carefully all but the bottom inch or so, avoiding sucking up any of the silt. A slight tilt to the storage vessel helps to get all possible water from it.

It is a bit tedious to filter the water prior to storage, but this will add to its eventual purity by removing much of the plankton and debris. A large filter funnel or a plastic sieve that fits over the top of the container in which the water will be stored can be lined with suitable filtering material such as filter floss or even several layers of clean cloth. If the filtered-off material is mainly plankton, feed it to your

treatment, usually by ultraviolet (UV) treatment in continuous flow setups. If you have abundant dollars, you may decide to install such equipment, whereupon you can run the newly collected water, after filtration, straight into the fish tank via a sterilization outfit.

Most authors recommend aeration of stored water before use, as it is likely to be severely depleted of oxygen and contain a lot of carbon dioxide. This is only likely to matter if a considerable

dilute the sea water you have collected, it can be heated to the right temperature by standing buckets or jerry cans full of it in a hot bath. With bath water initially at about 120°F and the seawater at 60°F, a rise to 80°F takes only 20 minutes or so. If you wish to dilute it a little, hot glass-distilled or demineralized water can be added directly—never use water from the hot water system as it is likely to be contaminated with metallic ions, particularly copper. A heater can

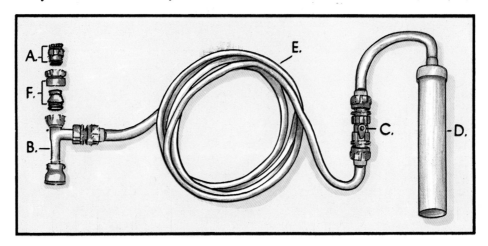

The No-Spill Clean and Fill by Python Products is available at most petshops. It enables you to empty or add water to any tank or pool. A) Two-piece Faucet Adapter. B) Faucet Pump. C) Open/Close Switch. D) Clear Gravel Tube. E) Clear, Flexible Tubing. F) Optional Snap Connector.

invertebrates! If you aim to use the water immediately, without storage, it is a very good idea to filter it, as you may remove parasites and even a proportion of smaller infective organisms. However, if it is for an invertebrate tank, don't bother, as the filter-feeders will have a ball. You may be introducing some galloping coelenterate pox, but in actual practice this doesn't seem to happen.

Methods have been described for the sterilization of new salt water, either physically or chemically. The artificial culture of larval forms of various fishes and invertebrates sometimes has been shown to benefit from such

change-over of water in the aquarium is contemplated.

I have never been conscious of any trouble when making a 25% change or less, but it would be wise to aerate for half a day if a bigger change is to be made or when setting up a new tank—which would normally be aerated itself before introducing anything living. Otherwise, by the time the water has been siphoned, carted around, and poured into the tank it will have been sufficiently aerated, even if it mattered, which it usually doesn't.

Remember that there may be a temperature difference between the aquarium and the stored water. If you do not wish to

of course be inserted into each bucket, but it takes longer that way.

Synthetic Mixes

The keeping of marine life has taken great strides since the availability of artificial sea water of reliable grades. Earlier products were often unsatisfactory and difficult to dissolve. Why this was so is difficult to explain, as a quite simple mix can support fishes satisfactorily for many months, but these early mixes did not even do that. However, all is now well, and any reputable product can be relied upon to support most marine creatures, including many invertebrates and algae. Some contain not only those elements known to be essential to fishes, invertebrates, and algae, but many that may possibly be so, although we are not sure. The proportions of some of the trace elements are different from those found naturally, sometimes to encourage algal growth, sometimes because they are present by accident but in amounts that do not seem to be harmful.

measure out for dissolving on each occasion, and you must therefore rely on the manufacturer having mixed his chemicals adequately. At least one very reputable mix is not even in composition, so that either a whole bag must be used or the unfortunate purchaser has to churn away at the mixture until he believes it to be uniform— if he knows about the problem! Otherwise, he may get a completely unbalanced mixture

It is quite wrong to assume that tap water suitable for human consumption is safe for use in making a salt water mix. It is not a question of hardness or pH or anything quite so simple, but of chemical treatment and sometimes even of metallic contamination thought to be harmless to us but definitely not so to fishes or invertebrates, particularly the latter. There are even cases known of contamination by industrial

To provide the proper medium for a reef tank the aquarist must take into account the variety of organisms that will have to live in it. The diversity of animals and plants in such a tank is great, and includes sponges, corals, worms, crustaceans, and fishes, to name a few. Each organism has its own particular needs with regard to chemicals, trace elements, clarity, density, etc., and if a vital component is missing it could die and cause havoc in the mini-ecosystem of the reef tank. Photo by John Burleson.

Synthetic mixes are usually available in the form of a plastic bag of powder to be dissolved directly in tap or other fresh water. Sometimes it already contains all the trace elements offered, sometimes a supplement of trace elements is to be added after the main bulk of material has dissolved. The bag is labelled as being for so many gallons or liters of eventual mix. It is advisable to purchase quantities that will supply immediate needs, so a whole bagful can be dissolved each time. Otherwise you must calculate how much to

by taking only a sample from the bag instead of using the whole lot. Fortunately, some manufacturers state that their mixes can be measured out as needed, so that worry is avoided. To approximate natural sea water, about 4½ oz. of an average mix per U.S. gallon of water is needed; fine adjustments can be made afterward if found necessary.

Contaminants

It is unfortunate that tap water is getting less and less suitable for immediate use in many districts.

wastes. The overall problem is fortunately smaller for the marine aquarist than for his freshwater counterpart, where hardness and pH matter, but luckily these factors are swamped by the addition of the synthetic salts— very hard tap water will be less than 0.1% salts; sea-water is 3.5%. The pH will be adjusted automatically whatever the starting value. Even chlorine doesn't matter much, as you will be aerating the mixture before adding it to the tank—if you don't, and you can get away with it in many instances, make sure

that there is little chlorine in the water or add a neutralizer (sodium thiosulphate, 1/2 grain per gallon).

The real problems are chloramines and metals. It is an increasingly common practice to add ammonia as well as chlorine to tap water to neutralize trihalomethanes, toxic compounds suspected of causing cancer, or to prolong the antibacterial action of the chlorine. The ammonia combines with any chlorine present to form chloramines, toxic compounds that cannot be blown off by aeration and must be broken down again to ammonia and chlorine, both of which must then be eliminated. This can be achieved by an excess of chlorine, an excess of sodium thiosulphate, or a commercial chlorine neutralizer. Twice the normal amount of sodium thiosulphate is often sufficient, namely one grain per gallon or 17 mg per liter. An excess of chlorine is not usually needed, but it can be supplied as sodium hypochlorite, 1 ml per gallon of a 5% solution. Sodium hypochlorite is the main constituent of most bleaches, but if you use a commercial product, be careful that it is the only chemical present and that no potentially dangerous substance is being added with it. The resulting ammonia and free chlorine can be removed by chemical absorbents (in the case of chlorine), by aeration, or by sodium thiosulphate. All of these measures must be taken before adding salts to the water. Commercial preparations to remove chloramines, with directions for use, are now available. Chloramines are highly toxic to fishes, as they pass readily through the gills and

(Above) A mature Emperor Angelfish, *Pomacanthus imperator*, because of its size needs a very large aquarium. Although larger tanks are more difficult to administer in many ways, it is generally easier to maintain the chemical balance in a larger tank than it is in a smaller tank. (Below) This chart indicates how much salt is required to change the specific gravity of water. At 25°C (77°F), for example, there is a 2 1/2% solution by weight to equal a specific gravity of 1.025.You can use a hydrometer to verify and constantly monitor the salt concentration (specific gravity) of your aquarium.

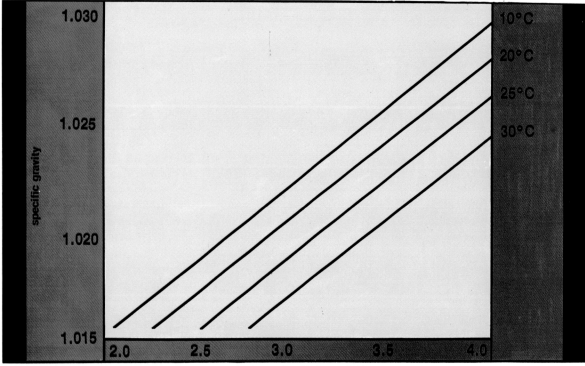

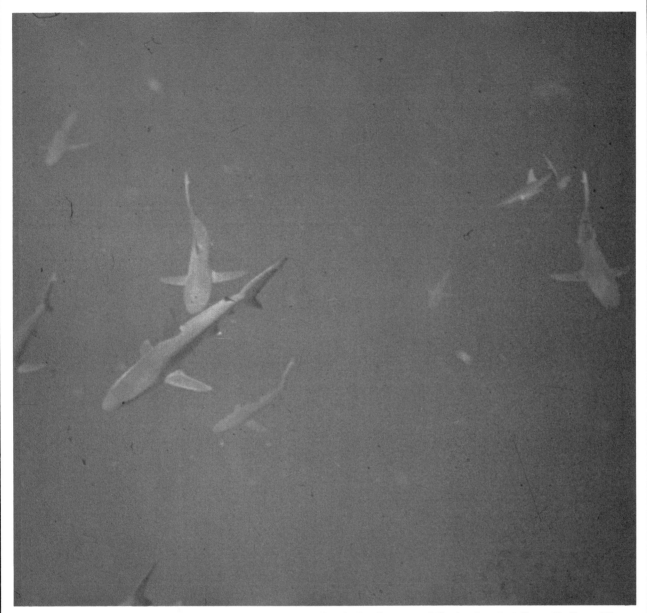

Sharks accumulating outside a typical harbor. They previously freely entered the harbor, but the pollution of the water has formed an invisible barrier through which the sharks will not enter. As civilization encroaches more and more upon pristine environments, fishes will have to adapt or disappear from the Earth...unless man wisely deals with the mounting problems of waste water, chemicals, gases, and solids, all of which eventually contaminate the sea.

combine with the hemoglobin in the blood to, like nitrites, make it incapable of carrying oxygen.

Metallic contamination is usually caused by new copper pipes, but sometimes it is caused by electrolytic or other breakdown of even old copper pipes, particularly in hot water systems. This is a reason for great care in using hot water from the tap to heat up or dilute natural or synthetic sea water. I have fallen into that trap myself and wondered why my precious invertebrates started dying. Metal storage tanks and piping other than copper can also be a source of danger. Some invertebrates are sensitive to very low concentrations of various metals. Copper, for instance, although needed physiologically by algae and crustaceans, is present in sea water at only around 0.5 microgram per liter (1 microgram is a thousandth of a mg), or in mixes up to 3 micrograms per liter, whereas even cold tap water can easily have 0.1 to 0.5 mg per liter, up to a thousand times the natural concentration. Fishes do not mind fairly high copper levels—indeed, it might cure them of some of the more common

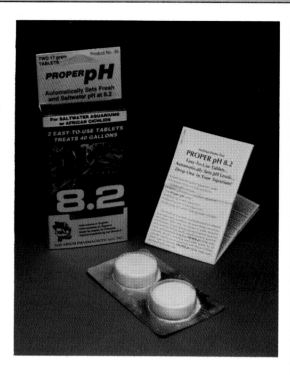

diseases—but even they should not be subjected to a constant high level.

Metallic contamination is hard to deal with at the low levels, chemically speaking, that we are considering. Carbon and some synthetic filter materials will lower gross amounts such as are used in temporary cures, but they do not get rid of all the metal. The same is true of any but an expensive ion exchange system. To guarantee a concentration of metals below 0.01 mg per liter (0.01 ppm) requires double passage through a series of ion exchange chambers. An alternative is triple-glass-distilled water, expensive in the quantities needed. Fortunately a copper level of around 0.02 to 0.05 mg per liter is tolerated by many invertebrates, but some species of anemones and practically all corals are affected by such levels, while nobody is quite sure about the long-term effects of any of them. Toxicity to algae also has to be considered. In my own experience anything over 0.05 mg per liter hits green algae and causes its replacement by red encrusting varieties.

(Above) To help maintain a proper pH in your tank manufacturers have developed pH tablets. These easy-to-use tablets automatically set the pH levels. (Below) Your local aquarium shop will be able to offer you many types of test kits designed to assist you in monitoring the quality of your aquarium environment. Photo courtesy of Aquarium Pharmaceuticals, Inc.

Once you have collected or made up the salt water to be used in a marine aquarium, its properties must be checked both at the start and periodically throughout the life of the tank. Underpopulated tanks with frequent partial water changes are likely to stay in order without much checking, but most marine aquaria should have regular measurements made to reassure the owner that all is well. The most frequent checks should be of pH and specific gravity. Other checks of nitrite and perhaps ammonia concentrations also should be made, frequently in a newly set up aquarium and less often later on. If copper is used in curing disease, its level should also be monitored.

Specific Gravity

The density of sea water is measured by a hydrometer, an instrument that compares the weight of a given volume of salt water with that of pure, distilled water. Pure water has a density of 1.0, typical sea water collected from the ocean a density of 1.025 at 60°F (approx. 15°C). Thus, whereas a liter of pure water weighs approximately 1000 gm at 60°F, a liter of ocean water will weigh 1025 gm. I say approximately, because the scientific definition of this ratio expresses it at the awkward temperature of 4°C (approx. 40°F), but most hydrometers are calibrated for 60°F; some, however, are now made for aquarists and are calibrated for 75° or 80°F. Make sure which you are buying.

In its usual form, a hydrometer is a heavy glass bulb with a narrow stem that projects vertically above the water. This stem bears graduations showing the specific gravity of the water in which the hydrometer floats. For measuring sea water, the marks will normally cover the range

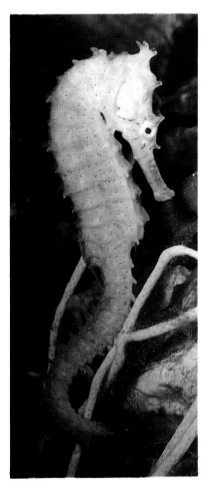

One of the most interesting of fishes generally recommended for beginners is the seahorse. It should be remembered that they require "holdfasts" for their tails, but they will make use of filter tubes, air lines, or even floating hydrometers for this purpose if nothing better is available. Photo of *Hippocampus erectus* by J. Kelly Giwojna.

1.000 to 1.050, from fresh water to a brine twice as dense as "normal" sea water. With an instrument calibrated for 60°F, a correction has to be made when other temperatures are concerned, a rise or fall of 10°F corresponding approximately to a fall or rise of 0.0015 in specific gravity when the true 60°F reading is 1.025. Thus, a reading of 1.022 at 80°F corresponds to one of 1.025 at 60°F. Note that in reading a hydrometer the eye

should be level with the water surface and the markings on the stem on the underside of the water surface should be read, not those in the air. A new type of hydrometer is immersed in the water and a moving pointer records the specific gravity, making life easier, as in the fish tank the usual instrument tends to bob about and be difficult to read; in fact, it is simpler to run off a sample into a measuring cylinder or jar and take the observation there.

There are other ways of measuring the density of sea water, but a hydrometer is much

A floating hydrometer, while old-fashioned, is still used by many aquarists. It is often very difficult to read.

the simplest. For great accuracy the direct method of drying off the salts and weighing them presents difficulties, because the salts are hygroscopic, meaning that they rapidly absorb moisture from the air and an exact weighing would have to be in a vacuum. Also, the temperature necessary to drive off the water initially decomposes carbonates, bicarbonates, some chlorides, and some bromides, with chlorine and bromine gases being set free.

Knudsen, in 1902, actually defined the *salinity* of water as the weight in grams *in vacuo* of solids dried at 480°C, with a correction for lost gases, but nobody does this in practice. Instead, electrical conductivity or the refractive index (light bending capacity) of the water is usually measured—or a hydrometer is used for an approximate reading.

Since the ratio of chlorides to other salts is practically constant whatever the actual specific gravity of a sample of sea water, the *chlorinity* may be measured and the salinity calculated from it. Measurement of the amount of chlorides present is very easy by precipitating them with a salt of silver, silver chloride having a very low solubility. Usually a solution of silver nitrate is run drop-wise into a measured sample of the water to be titrated and some potassium chromate is added to the sample. When the last of the chlorides are precipitated as silver chloride, the silver will then combine with chromate to give a red precipitate, indicating the end of the titration.

The following table enables you to convert salinity to specific gravity at 60°F, with corrections for other temperatures. It can be used in making up synthetic salts, for instance, the percent salinity giving a guide to the amount to use to achieve a desired specific gravity—most simply in grams per liter, but the table also gives an oz. per gallon equivalent. When salinity is expressed as a whole number (25, 30, 35) it refers to parts per thousands and may be derived by moving the % salinity decimal one place to the right. Thus, % salinity 3.11 = 31.1 ppt.

Table 1 — The relation between salinity (% salts) and specific gravity at different temperatures

Specific gravity at °F

68°	65°	70°	75°	80°	85°	Salinity %	oz /gal.
1.0150	1.0145	1.0138	1.0130	1.0124	1.0115	2.06	2.75
1.0160	1.0154	1.0148	1.0140	1.0134	1.0125	2.20	2.94
1.0170	1.0164	1.0158	1.0150	1.0143	1.0134	2.33	3.11
1.0180	1.0174	1.0168	1.0160	1.0153	1.0144	2.46	3.29
1.0190	1.0184	1.0178	1.0170	1.0163	1.0154	2.59	3.46
1.0200	1.0194	1.0187	1.0180	1.0173	1.0163	2.72	3.64
1.0210	1.0204	1.0197	1.0189	1.0183	1.0173	2.85	3.81
1.0220	1.0214	1.0207	1.0199	1.0192	1.0183	2.98	3.98
1.0230	1.0224	1.0217	1.0209	1.0202	1.0193	3.11	4.16
1.0240	1.0234	1.0227	1.0219	1.0212	1.0203	3.24	4.33
1.0250	1.0244	1.0237	1.0229	1.0222	1.0212	3.37	4.51
1.0260	1.0254	1.0247	1.0238	1.0231	1.0222	3.50	4.68
1.0270	1.0264	1.0256	1.0248	1.0241	1.0232	3.63	4.85
1.0280	1.0274	1.0266	1.0258	1.0251	1.0241	3.76	5.03
1.0290	1.0284	1.0276	1.0268	1.0260	1.0251	3.89	5.20
1.0300	1.0294	1.0286	1.0277	1.0270	1.0260	4.02	5.37

In practice, it usually is not easy to read a hygrometer to the fourth decimal place given in the table, so if your instrument reads 1.023 at 78°F look down the 80°F column to 1.0231 and note that the 75°F column says 1.0238. At 78° the expected reading would be 1.0238—(7x3)/5 = 1.0234, indicating a reading at 60°F of a little over 1.0260, somewhat high. Why –(7x3)/5? Because the difference between 1.0238 and 1.0231 is 7, and we need to go 3

Crabs, like this Flammulated Box Crab (*Calappa flammea*), are common inhabitants of invertebrate tanks. Their powerful claws can crush the shells of marine snails, so their tank-mates must be selected with care. Photo by U. Erich Friese.

of the 5 steps between them to 78°F. Actually, it is good enough to take the nearest column, as we did to start with, and read off from that—we still get a little over 1.0260.

Most marine aquarists agree that a slight dilution compared with natural sea water is best for fishes, although not necessarily for invertebrates, particularly soft-bodied ones like coral. Most others can stand mildly diluted water, but it is not so much to their advantage as it is for fishes. This is because a lower salinity slows down a fish's metabolism, allows easier excretion of soluble wastes, and permits a higher oxygen concentration in the water, since oxygen is less soluble in salt water than in fresh water. As good oxygenation is more essential to sea fishes than to fresh water fishes, this is of importance. Anything down to 1.017 is recommended, and for fishes alone this is fine. In fact, one recent writer states that he keeps his fishes at 1.013, with perfect success and a lower incidence of disease. The big difficulty accompanying any serious dilution is that your dealer and the ocean don't keep their fishes at 1.017 or 1.013, so they must be acclimated rather slowly before being placed in the aquarium. Steps of perhaps 0.002 with a few days between each change would be a sensible precaution to be taken during a quarantine period.

Hydrogen Ion Concentration

This is pH, or $-\log_{10}H^+$, a measure of the concentration of positively charged, acidic, hydrogen ions in solution. A low pH means acidity, pH 7 is neutral, and a high pH means alkaline conditions. In dealing with aquaria, we are interested in chemically very mild departures

from pH 7, but they are important to the plants and animals. Even a mild acid like vinegar is at pH 3 or less, while a strong acid is around pH 1. A mild alkali like sodium bicarbonate in solution is at pH 9.6 approx., but a strong alkali like caustic soda is over pH 13. Sea water is at pH 8.0 to 8.3 in the open ocean, but in deep water or when mildly polluted it may fall to as low as pH 7.6. In the aquarium greater falls tend to occur and must be prevented. Most marine aquaria are found to be between pH 7.6 and 8.3, but it is best to keep them above pH 7.8, and preferably above pH 8.0 if containing most types of invertebrates. A tank may exceptionally rise above 8.3 as a result of over-oxygenation, usually due to algae in sunlight producing visible bubbles of oxygen and mopping up the carbon dioxide which lowers the pH. This doesn't matter in a well-kept tank but could release ammonia in a polluted one.

The pH of an aquarium can be measured in various ways: electronically, or by chemical indicators on paper, or by chemical indicators in solution.

Keeping invertebrates and fishes in the same aquarium can be very rewarding. Photo courtesy Midori Shobo.

The electronic pH meter is unnecessarily expensive but very accurate. pH papers to be dipped in the water and compared with a color chart tend to be inaccurate to a dangerous degree, so that

liquid color comparison is by far the best. A sample of water is taken in a small tube and a stated number of drops of indicator as directed by the manufacturer is added. The sample is mixed well and compared for color with a chart (least recommended), colored glasses, tubes of liquid (quite good as long as they do not fade with keeping), or, best of all, with colored glasses through which another untreated sample of the tank water is viewed so as to correct for any color in the tank water itself. In a typical comparator with phenol red as the indicator, at pH 7.4 the color is yellow, changing through pinkish to light red at 8.0 and to very dark red at 8.4.

Sea water at normal pH values has a weak buffering capacity, which means that it can withstand small amounts of acid or alkali without change, or much change, in pH. This capacity is not as strong as some authors assert, and in the aquarium it is weaker than in the ocean and must be maintained by one means or another. As the main tendency is to acid conditions, plenty of alkaline gravel such as coral sand or shell grit is indicated, with adequate water changes and the periodic addition of an alkali such as sodium bicarbonate. Some manufacturers sell sodium carbonate or even caustic soda in solution to be added drop-wise to the water, but the bicarbonate, although needed in greater quantity, gives a better buffering action and is gentler in use. A fish happening to get a mouthful of caustic soda will be decidedly unhappy.

The main buffers in sea water are carbonates and bicarbonates, with the latter the dominant salt. They repress fluctuations in pH due to changes in carbon dioxide concentration, the primary cause

Aulonocara nyassae, a cichlid from Lake Malawi, does best in an alkaline environment with a pH around 8. The fish shown below are *Capoeta hulstaerti* from Zaire. They thrive only in a very acid water with a pH around 3. Both are freshwater fishes. Of the thousands of freshwater fishes, most have definite pH requirements from pH3 to pH8. Marine fishes almost all have the same pH requirement of about 8 (actually a range from 7.6 to 8.2 is acceptable).

of acidity. The reaction is, in terms of ions in solution:

$$2HCO_3^- \leftrightarrow CO_2 + H_2O + CO_3^{--}$$

If carbon dioxide, CO_2, is added the equation moves toward the left and helps mop out the gas as bicarbonate ions.

At temperatures around 77°F (25°C), typical of a tropical marine aquarium, it requires about 0.4g of sodium bicarbonate per gallon to raise the pH from 7.6 to 8.2, the optimal pH. This is roughly a level teaspoon per 10 gallons. To raise it from 7.8 to 8.2 needs about 0.3g per gallon, and from 8.0 to 8.2 only 0.2g per gallon. Lower temperatures increase these requirements a little, by around 20%. If your tank is as low as pH 7.6, don't raise the pH all at once; make a gradual adjustment over 2 days. The necessary amount of bicarbonate is best dissolved first in a little tank water. It is very rarely necessary to lower the pH of a marine aquarium, but this may best be done by adding acid sodium phosphate (NaH_2PO_4), say a teaspoon per 20 gallons and then testing the result.

Ammonia, Nitrites and Nitrates

These products of bacterial breakdown of nitrogen-containing organic substances must be kept to reasonable levels. The actual ppm (parts per million) as measured by various kits is not always expressed in equivalent nitrogen concentration, the usual procedure, but kits do usually give the safe levels on their own scales. Nitrites are determined, for instance, as nitrite nitrogen-N, and to obtain the equivalent weight of nitrite ions, NO_2, we must multiply by 3.3.

Ammonia measurement by chemical means is so inaccurate that it is not recommended. Only the ion electrode method can be relied upon to give accurate results. At most 0.1 ppm as ammonia-N should be tolerated. Luckily, it is normally suffcient to measure nitrites, or rather nitrite-N, for which some commercial kits are reasonably accurate. Even the best kits degenerate with keeping, indicated by cloudy or precipitated test solutions, which must then be discarded. In an established tank, if nitrites are low ammonia can be assumed to be low also. A level of 0.25 ppm would be a conservative, safe figure, but many fishes seem to tolerate several times this level for at least some time. Nitrates are measured, also as nitrate-N, by kits similar to the nitrite ones, and the same warning about degeneration applies. Nitrates are far less toxic than ammonia or nitrites. Up to 40 ppm is usually regarded as safe, and fishes in some public aquaria have been found to tolerate much higher levels, if gradually built up.

Dissolved Gases

The important dissolved gases

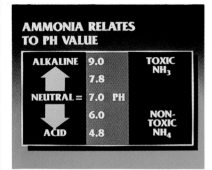

AMMONIA RELATES TO PH VALUE

ALKALINE	9.0		TOXIC NH₃
	7.8		
NEUTRAL =	7.0	PH	
	6.0		NON-TOXIC NH₄
ACID	4.8		

the top of the mine shaft.

Nitrogen is rarely of importance, but in a supersaturated state it can cause the "bends", even in fishes. This can occur if water is delivered under pressure in public aquaria, but it rarely happens in the home.

are oxygen and carbon dioxide. These can be measured with expensive kits on the market, but it is rarely done. Sea water accepts less oxygen than fresh water, and the higher the temperature the lower the solubility of the gas. Thus, at 60° F, fresh water can dissolve 10.1 ppm of oxygen and sea water only 8.2 ppm; at 80° F fresh water dissolves a maximum of 8.0 ppm and sea water only 6.5 ppm. At the surface of the ocean at 60°F, dissolved oxygen may reach 7.0 ppm, but in the tropics, at 80°F or above, 4.0 ppm would be a common reading. Tropical fishes can stand such a low oxygen level, but coldwater fishes cannot. Even tropicals are unhappy much below 4.0 ppm, so the oxygen level in the aquarium must be kept as near to saturation as we can manage. We are helped by the rapid solubility of oxygen, so good aeration and water circulation are sufficient to keep things in good order.

In contrast to oxygen, carbon dioxide moves only slowly into or out of water. In a totally covered tank it can accumulate over the surface of the water and cause trouble. A good flow of air, via aeration or a brisk return of filtered water at the surface, helps to avoid this and also helps to avoid overtaxing the rather weak buffering capacity of sea water at normal pH values. Carbon dioxide is a heavy gas and kills the miner's dog without harming the miner, who breathes air from

Copper

Copper is so frequently used to cure some common fish diseases that its measurement when so used is highly desirable. It is unfortunate that many commercial cures containing copper do so in chelated form, in which the metal is strongly bound to a chemical "carrier" and slowly released. No sure method is available for measuring copper levels in such circumstances, and the following applies only to copper added as a simple salt, the sulphate or citrate. Copper may also be present in unwanted quantities in tap water, which should be tested to see if it is suitable for use as

(Center and right columns) Tube worms are spectacular additions to any miniature reef tank. It is surprising how many will turn up on living rock after a time. These photos give only a small indication of their variety. They filter small organisms from the water so require a current that can carry the tiny food particles to them. Photos by Courtney Platt and Mike Mesgleski.

make-up water with synthetic mixes. My own tap water measures 0.41 ppm metallic copper, a higher level than is usually recommended for the cure of disease!

Copper kits are of variable reliability—as with ammonia and nitrites, it is just darned difficult to measure the very low amounts we are considering. A chemical laboratory would usually concentrate the sample before making a measurement or would use a method such as atomic absorption spectroscopy, which isn't possible for most of us. So take advice from your dealer or fellow aquarists before choosing a kit. Copper is highly toxic to most invertebrates and toxic to marine fishes at levels above 0.3 to 0.4 ppm metallic copper (i.e., not copper sulphate, $CuSO_4 \cdot 5H_2O$ which weighs approximately five times its copper content). In an invertebrate tank, copper must be kept virtually unmeasurable for continued success with most species, whereas with fishes alone some aquarists and public aquaria keep a level of around 0.15 ppm constantly in the water as a precautionary measure.

The marine aquarium is for the more sophisticated aquarist. Usually it is not for children as the water chemistry and fish behavior require a fairly thorough education in fundamentals (that's what this book is all about). Aside from the mental stimulation, the marine aquarium can be a strikingly beautiful ornament as shown by this Aqua-Module aquarium.

The marine aquarium needs adequate lighting, heating (rarely cooling), aeration, and filtration. We shall not be discussing the "natural" system, in which filtration is omitted, in this chapter.

Lighting

The intensity of lighting required in a tank depends on the nature of its contents. Fishes need enough light to see and be seen by; algae of a desirable nature need considerably more light to flourish; tropical green algae, corals, and some anemones need quite intense illumination. In many circumstances the choice of lighting goes to fluorescent tubes, because they give off less heat than other types of lamps and offer a wide choice of spectra (distribution of light across the different colors).

The intensity of light in the sea depends on the depth of water and its turbidity, but in the aquarium we aim for such purity that for all practical purposes the light is unabsorbed by the water, which is never deep enough for such factors to take effect. If the tank is neglected and the yellowing or greening of water occurs, in deep tanks the loss of light may be noticeable and affect algal growth.

In the sea, in reasonably clear water, about 75% total radiation from the sun is available at 3-4 ft below the surface, only 50% at 6-8 ft, 20% at 15-20 ft, and so on. Moreover, the nature of the radiation changes, since the red end of the spectrum is absorbed almost completely in the first 15 feet, while the blue end penetrates much deeper. The apparent colors of undersea creatures thus change dramatically as we descend. Red crabs and shrimps look black at depths because no red light

reaches them—an explanation of why they can afford to be red! Green algae, which need much light, can only grow near the surface, while those corals that include them as symbionts living in their tissues must not be at too great a depth either. Green algae look green because they absorb red and blue light and only reflect the greens. While red and brown algae also do this, other pigments mask the green color and also help to absorb light for photosynthesis (the manufacture of sugars from carbon dioxide and water).

The early fluorescent tubes gave off isolated bands of different colors, single wavelengths scattered across the spectrum instead of the continuous spectrum of natural sunlight or incandescent lights. Modern tubes have overcome this to a large extent but still do not give an even spectrum, having humps and bumps of different colors. None of this matters as long as their effect is pleasing and the plants, if any, can get adequate light for photosynthesis. The typical plant growth bulbs,

for example, have big peaks in the blue and red, just where needed for plant growth, whereas a typical "daylight" tube has plenty of blue but very little red, only 5% or so of its output. A good choice for both fishes and plants is a "warm white" or "natural" tube, both giving a look of white light but with more adequate red content. From the viewer's standpoint, plant growth tubes, of which there are several types, also enhance the colors of fishes and are preferred for that reason by many aquarists, plants or no plants.

The spectrum of a fluorescent lamp is one thing, but its total output is another. Standard plant lamps put out 23 lumens (a lumen is a standard of light intensity) per watt, daylight 64, and warm white 81 lumens. Thus, warm white tubes give about 3½ times the total illumination of standard plant growth tubes. Although plant tubes put out 40% of their energy in the red and 27% in the blue, warm whites with their output of 12% in the red and 13% in the blue actually give as much red

The intensity and quality of light required for a given aquarium vary with the types of inhabitants and the desired effects. These tubes here have been specially developed to illuminate reef aquaria, providing the necessary intensities and wave lengths. Photo courtesy of Coralife/Energy Savers.

An example of halide lights housed in an aquarium reflector.

and more blue total than plant bulbs, plus higher middle frequencies. The new actinic-03 tubes give a peak of light in the blue and have been found to be particularly effective in maintaining corals. Plant tubes look purplish because they are so deficient in the yellows and orange compared with the "whites" of any kind. It's really up to the aquarist to choose what he wants to stimulate and see. Warm white tubes will grow algae at least as well as standard plant growth types (there is now, however, a wide-spectrum plant bulb that more closely resembles a warm white in output), but plant growth types enhance the look of the tank and fishes.

Top lights for an aquarium are usually enclosed in a reflecting hood to increase the amount of light reaching the tank and to hide the bulbs from view. Beneath them will usually be glass covers, but these are omitted on miniature reef tanks to increase oxygen circulation. A typical fluorescent tube rates 10 watts per foot; a warm white gives 81 lumens per watt, or 400 lumens per foot of light into the aquarium (this is about half of its total output). If this falls on a square foot of tank, then at least at the surface this is 400 lumens

per square foot of surface area, less at any depth but not vastly less, as internal reflection gives a better figure than one might expect. Tropical sunlight can provide up to 10,000 lumens per square foot. Higher green algae need about 1,600 to flourish; higher red and brown algae need 200 to 1,000; simple unicellular algae around 400. To grow nice, lush green *Caulerpa* or other fronded algae we thus need more than one tube per tank, considerably more in large tanks. Over a 2-foot tank, two tubes the length of the tank could suffice, but three would be better; over a 3-foot or bigger tank, at least four tubes are needed (if you wish to keep plants). To just see fishes, etc., adequately, one tube is quite enough.

If you are planning a clean, relatively algae-free tank that you will keep bright and nice-looking by frequent bleaching of coral, one fluorescent tube is fine. It will tend to produce red or brown algal growth, not very nice-looking, but you are going to take care of that. If you want to grow green algae, particularly the higher, fronded forms, more light is needed, 2 or 3 warm white tubes over a 2-foot tank, 4 or even more over a 3- or 4-foot or larger tank—which will usually not be deeper than 2 feet, only longer. This may be difficult to provide unless the aquarium is fairly wide, and you may have to have a hood specially made with the tubes closer together than the usual arrangement. The same requirements apply to corals and many anemones, which depend on the algae living within their tissues to thrive.

Now that there is more interest

in miniature reef aquaria that are sophisticated versions of the natural system, lighting by mercury vapor or metal halide lamps is being explored. These must be hung above the aquarium because of their heat production. Metal halides seem particularly useful and can provide an intense, natural-looking light.

The lights on an aquarium should be on for at least 12 hours per day, and 14 hours is not excessive. A timer switch to deal with this is often recommended, but it is only advisable if the room in which the aquarium sits is in daylight or artificial light *before* and *after* the tank lights go on and off. This is because fishes respond with panic far too often if they are suddenly illuminated or suddenly plunged into darkness. In addition, marine fishes need time to find their usual resting places at nightfall and are much put out if not given it. As you will probably wish to enjoy your aquarium in the evening, a schedule like 8 a.m. to 10 p.m. is suitable; even 8 a.m.to 12 p.m. would not be unreasonable for night owls, and the fishes will accommodate themselves. Keep a regular schedule and you will see many of the fishes settling themselves down at the appropriate time.

Heating

Care must be taken with heating a marine aquarium, as salt water has an enormous capacity to creep everywhere.

The ocean waters are filled with interesting plants and many of them might thrive under aquarium conditions. This green marine alga was collected by Dr. Herbert R. Axelrod. He maintained it for years in a very strongly lit aquarium.

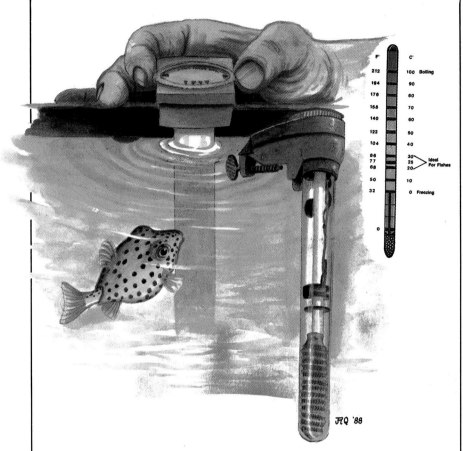

Modern heaters for small aquaria are of glass or plastic exterior with a rubber or water-resistant bung of some kind through which the leads enter; there may be an external temperature control knob or the control may be inside. Most modern heaters are combined with a thermostat and so have this traditional control arrangement. Otherwise, the thermostat is a separate instrument and can be situated outside the tank, with a sensor applied to the tank externally or internally. For ease of operation, the combination instrument is usually preferred.

The best instrument for a marine tank is a submersible heater-thermostat combination *guaranteed* to be suitable for salt water. It avoids the danger of a non-submersible heater being accidentally submerged or penetrated by splash, drip, or

Most heaters are adjustable from the outside, such as is shown in the drawing. The ideal temperature for most reef fishes lies between 20-30°C or 68-86°F.

creep. If laid at the bottom of the tank, it also avoids the danger of our forgetting to switch it off when making water changes and shattering a hot glass heater when filling up again. An equally good alternative is an external thermostat with a submersible heater or heaters.

Control of tank temperature to ± 1.5°F is adequate and avoids too frequent an operation of the thermostat, which if of the usual bimetallic strip construction tends to arc and wear with constant use. Practically all our usual marine aquarium fishes are accustomed to some temperature swing, as long as it is not sudden and not too extreme. It is best to

purchase a heater that adequately controls temperature without having too great a rating—this is because it will then be incapable of cooking the fishes if the thermostat should stick in the "on" position. A 20°F margin is ample in practically all normal conditions. Such a heater permanently on will push the temperature to 20°F over room temperature, so if you have permanent air conditioning set for 75° or 80°F, keep the heater down to a 10°F margin or do without tank heating at all.

The smaller the tank, the more heat is required per gallon of water, since heat is lost more rapidly from the greater surface area per gallon. As a guide, Table 2 gives the heater capacity in watts required to keep various normally shaped tanks (an approximate double-cube construction) 20°F above room temperature. Reasonable departures from a double cube don't make any appreciable difference—i.e., a 40″ x 18″ x 22″ tank wouldn't significantly differ in heat loss from a 40″ x 20″ x 20″ tank. Even long tanks such as 72″ x 24″ x 24″ will not differ enough to matter. It is the very oddly shaped ones that would need special calculations of their own.

The calculations are based on the formula

$$\text{Watts per gallon} = t^2 l_{24} / 100 \, l_x$$

where t is the maximum temperature differential, l_{24} is the length of a typical 15-gallon tank (24″) and l_x is the length of the tank in question. The wattage in the last column is the ideal one; naturally you will have to select the nearest available heater in actuality—take the next one above the calculated rating needed rather than the one below it if it is much below.

Table 2 — The Theoretical Wattage of Heaters for Different Sized Tanks			
Tank length *in inches*	Gallon *capacity*	Watts *per gallon*	Total *Watts*
12	1.9	8.0	15
18	6.3	5.3	33
24	15.0	4.0	60
30	30	3.2	96
36	50	2.6	136
42	80	2.3	183
48	120	2.0	240
Above 48*	-	2.0	-

*A tank over 48 inches long usually will not be more than 24 inches deep, so the watts per gallon stay at 2.0 approx.

In Table 2, the 12″ and 18″ tanks are included more to show how the necessary wattage increases than in the expectation that such small marine tanks will usually be kept.

A thermometer is a must in a tropical tank and is advisable in any tank. Two types are suitable—an alcohol thermometer for inside use and a stick-on type for application to the outside of the tank. The latter has temperature-sensitive liquid crystals that light up the appropriate little compartment showing what is the temperature. They usually give a reading slightly influenced by the room temperature, but not enough to matter. Mercury thermometers are best avoided as a breakage may release a toxic amount of the metal into the aquarium. Clock-type thermometers are not always reliable and tend to come unstuck from the glass. Any thermometer should be checked against a known reliable one as they are liable to be several degrees in error. Finally, no metallic parts—clips, backing, or anything, can be placed in a saltwater tank.

Aeration

Aeration in a marine aquarium will often be part and parcel of filtration, but additional aeration may be needed if the filtering arrangements do not stir the water surface adequately, the tank is overcrowded, it is overheated, or it is in need of a clean-up. It is always advisable to have an airstone or two that can be put into operation rapidly if needed. The point about aeration that is most frequently misunderstood is that not much gas exchange occurs between the bubbles and the water unless there is a very great number of tiny bubbles. Indeed, the main function of aeration is to stir the water, bringing lower levels up to the surface for exchange with the atmosphere. This has been demonstrated by using nitrogen or even carbon dioxide (the latter in freshwater tanks) to effect a good exchange.

The stirring effect is best achieved with bubbles around 1/30th to 1/50th inch in diameter; larger ones do not carry so much water up with them and smaller ones dissipate as a fine mist.

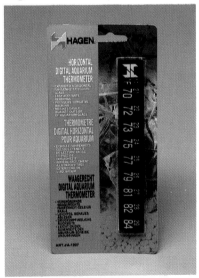

The easiest aquarium thermometers to read are the digital thermometers that can be mounted on the outside of the aquarium glass. They are slightly influenced by room temperature but this is for the most part negligible. Photo courtesy of Hagen.

A group of *Dascyllus trimaculatus* have made their home in an anemone, a scene than can be reproduced in your reef aquarium. Photo by Douglas Faulkner.

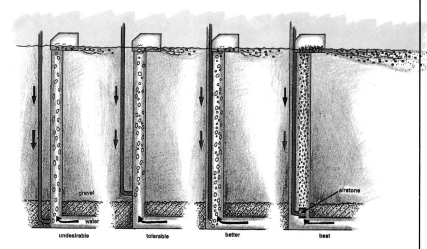

(Above) A Yellowhead Jawfish with a batch of eggs in its mouth. A deep layer of gravel and sand is needed for this fish. Photo by Cathy Church. (Below) Submersible heater/thermostat combinations are available at your pet shop in various wattages. Photo courtesy of Hagen.

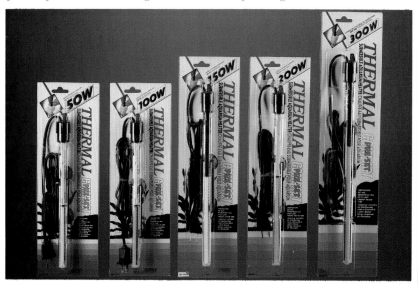

Maximally effective aeration can be achieved with quite a modest stream of bubbles, but in the marine aquarium it does not carry with it a doubling of fish capacity as in the freshwater tank. This is because marine fishes need almost as much oxygen as they can possibly get and the function of aeration in one form or another is to see that they get it. The water movement involved also helps the fishes, accustomed to moving waters in their natural state. You will probably notice that an airstone gives finer bubbles in salt water than in fresh water, a useful happening with most stones. The best stones for marine use are of fused small-grain ceramic or similar inert materials. Wooden devices give fine bubbles but tend to clog frequently.

Air pumps come in any desired size. Modern diaphragm pumps are quiet, reliable, and relatively inexpensive. Only an advanced fancier or breeder with many tanks need think of anything else. Gang-valves and connectors of various types make distribution of the air to filters and airstones very easy, but be sure to use soft, pliable tubing that does what you want and not what it wants. Make sure that there is no undue resistance from any equipment, particularly airstones, as this either causes pump damage or causes connections to blow. You should be able to blow easily through whatever the air-line leads to, or it is offering too much resistance. About 4 lbs per square inch is the maximum desirable pressure, about 1/4 atmosphere.

Filtration

There are about as many varieties of filter as there are manufacturers of them. Generally speaking, some form or other of filtration is mandatory in a marine aquarium—*except* in the "natural" system. It pays before going into details to consider what a filter can do to the benefit of an aquarium. It can:

1. Remove particles of uneaten food, fish feces, and any other debris from the water so as to keep it unclouded and "sweet".

2. Remove floating algae and even bacteria if it is fine enough.

3. Remove coloring matter if it is chemically active as well as having a purely mechanical action.

4. Remove or inactivate other chemicals such as the invisible dissolved products of protein breakdown (ammonia and nitrites).

5. Help maintain a suitable pH.

6. Remove chemicals or medicines after they have done their job—if they did! Also it can remove breakdown products of such substances that may be poisonous.

Filters were initially installed with functions 1 and 2 above in view. In large, usually public aquaria they were deep beds of sand or similar fine material that would hold back any gross

particulate matter. Settling tanks might also have been employed for the same purpose. In smaller home aquaria, sand is usually replaced by filter floss, loose or in pad form, and to it usually will be added finely divided activated charcoal (carbon) and possibly crushed coral or other limey material designed to help keep the pH right. It is doubtful how much the limey materials help at pH 8.0-8.3, however. The additions introduce functions 3, 5, and 6. If left long enough for bacteria to grow on the filter material, an appreciable degree of

Power heads are compact but relatively efficient water pumps. They are very good at creating water movement and aeration. One of their most important applications in the marine aquarium is to serve as a source of power for greatly increasing the flow of water through an undergravel filter, thereby providing a greatly increased supply of oxygen to the beneficial bacteria in the filter. Photo courtesy of Hagen.

Cannister filters are very powerful, though cleaning them is far from an enjoyable task. They have three filtering chambers and come in many different sizes.

function 4 begins to occur, as these bacteria will convert ammonia and nitrites to nitrates. This assumes that the filter is large enough for adequate bacterial growth, small filters will need cleaning out so often that they don't allow much growth to occur, and even if it does it won't be of much use.

Although filters in freshwater aquaria are often placed inside the tank, an adequate filter for a marine aquarium needs to be external to it, unless it is merely a carbon filter auxiliary to an undergravel (biological) filter that does most of the work. In the absence of an undergravel filter, the external filter should be on the order of 10% of the tank capacity. Then it eventually becomes an external biological filter and should be left undisturbed for months on end and never completely renewed. However, an undergravel filter is so much better in ordinary circumstances that it is far to be preferred unless you go to the trouble of adding a dry-wet external filter as in the miniature reef aquaria.

Carbon Filters

Very fine-grained activated carbon, *not* the filter "coals" still available, is capable of an enormous uptake of a great variety of materials. The method of preparing it produces pores within the grains that have an extraordinary total surface area per weight of carbon. Production involves heating the carbon to 1650°F, which is why it is not feasible to try to reactivate used material. The advice often given to bake in the oven or to treat with steam is misguided, as it does nothing.

The very large surface of high-grade activated carbon can absorb up to 50% of the weight of the carbon itself of various gases;

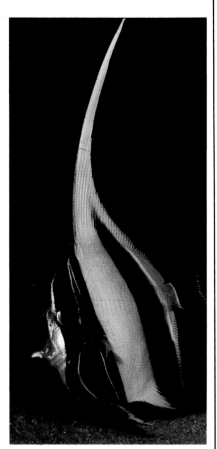

The Moorish Idol (*Zanclus cornutus*) has been considered a delicate species by many aquarists. It requires the best of water conditions. Photo by Karl Knaack.

coloring matters including methylene blue; heavy metals such as copper and zinc; organic materials such as sugars, amino acids, and some proteins; and many antibiotics and other "cures". It does not absorb significant amounts of ammonia or nitrites that are dealt with by a biological filter until it becomes one itself by bacterial growth— but by then it will have lost much of its capacity to remove other

(Above) Aquarists have kept Anemonefishes, like this *Amphiprion ocellaris* with their anemones for many years. Photo by Burkhard Kahl.

materials. Activated carbon also tends to lower the pH and remove vitamins and trace elements and is therefore not an unmixed blessing.

It is best to wash the carbon with a little sea water before use to remove unsightly black dust that may get into the tank otherwise and to leach out any toxins it may contain. Very high-grade carbon intended for chemical use is safe with a brief wash, but commercial grades need preferably up to 24 hours of circulating saltwater washing. Only a few ounces per 20 or 30 gallons of aquarium water are needed, when its effects on pH are minimal, and it can be used for several months, even up to a year in an understocked, clean tank. In fact, as its pores become somewhat clogged by bacterial growth and it no longer removes trace elements so effectively, it is better for the fishes and particularly for invertebrates than was the raw material. There is of course a limit to its useful life, after which it becomes only a small, rather ineffective biological filter. There are also limits to the capacity of carbon to remove various chemicals. It reduces their concentration to a varying degree but doesn't get rid of everything completely. Thus copper is reduced to below 0.1 ppm, but only slowly beyond this and never totally; some invertebrates cannot tolerate a tank cleaned up after copper treatment. For the same reason, not all of a trace element is removed—luckily! When buying activated carbon, make sure that it is in a sealed container and keep it tightly closed if you use only part of it. Otherwise it will pick up volatile materials such as paint fumes and fly spray from the air, which may be toxic.

Special disposable filter cartridges are available that fit certain filtering devices. Consult with your petshop owner as to which system is best for you.

Certain filter media help to purify the water in your tank, removing many toxic or other undesirable materials. Photo courtesy of Boyd Enterprises.

Ion Exchange Resins

There has been a great deal of controversy about the value of ion exchange resins in the marine aquarium. Their use in fresh water to remove heavy metals and to soften it is not in question, but exactly what they do to sea water is another matter. A typical ion exchange resin, in the sodium phase, will take up any metal heavier than sodium and release a corresponding amount of sodium into the water. Hard water, containing calcium, will finish up with much less calcium and some sodium instead, which will be harmless. In the acid phase, hydrogen is released and any metal, including sodium, is taken up. As this usually acidifies the water, the acid phase normally is not used in aquaria. In sea water, there is so much calcium and magnesium present that such a resin becomes rapidly exhausted as far as those metals are concerned, but it will still exchange these for even heavier metals like copper and presumably many trace elements. It can be expected to clean up after copper treatment, but what else it does is mostly conjectural. However, vitamins, coloring matter, and other organics will not be affected, and the obvious use of ion exchange resins is to clean up after copper treatment or to remove excesses of any heavy metal. There are, however, ammonia-removing resins available that can pick up such materials.

Some of the small shrimp make interesting companions for sea anemones. This is *Periclimenes brevicarpalis*. Be sure to check whether or not the animals proposed for your tank are compatible. Photo by Aaron Norman.

Water Polishers

Fine cloudiness of water may sometimes be a real problem, particularly in a newly set up tank. It may be bacterial or result from inadequate washing of the substrate or even carbon, and be very difficult to clear up by the usual methods. A diatomaceous earth filter, which has a capsule of very fine siliceous diatom shells, will do it but requires a powerful water pump to force the water through the filter. There are available products that cause clumping of the fine suspended particles so that they no longer pass through an ordinary filter, even an undergravel filter. Quite harmless in proper dosages, such a substance can be a boon.

Biological Filters

Although already mentioned, as regards filtration we have left the best until last. A biological filter is any device in which a

massive culture of bacteria is arranged to deal with the conversion of the toxic products of protein breakdown, which constitute the main hazard in the marine aquarium. The most toxic of these is ammonia, and it is particularly poisonous in the marine tank because of the high pH of sea water. Dissolved ammonia takes the following forms in water:

$$NH_4OH \longleftrightarrow NH_4^+ + OH^- \longleftrightarrow NH_3 + H_2O$$

Ammonium hydroxide $\longleftrightarrow$ Ammonium and hydroxyl ions $\longleftrightarrow$ ammonia and water

The higher the pH, the more the equation is pushed towards the right-hand side, resulting in the freeing of more ammonia gas, the most lethal form. An increase of one unit of pH results in approximately ten times the

amount of free ammonia, so that an aquarium at pH 8.3 has much more present than one at pH 7, typical of fresh water. That is why control of waste products, especially ammonia, is so important in the marine aquarium. There are small temperature and salinity effects, but nothing to compare with the effect of pH.

The effect on free ammonia of different pH and temperature values is shown in Table 3.

Surgeonfishes are normally nibblers. This means that to keep a Blue Tang (*Acanthurus coeruleus*) like this one you must (1) provide the proper vegetable matter for food, and (2) make sure that the algae you want to keep is not on its diet. Photo by Mike Mesgleski.

Ammonia is part of the nitrogen cycle, which describes the natural history of nitrogen-containing compounds in the aquarium, in the soil, and elsewhere. Such compounds, mainly proteins, are an essential in the diet of all animals and so are pumped into the aquarium constantly in feeding. They are released after breakdown as fish feces, and also by any decay of plant life or uneaten food. The main product of breakdown, ammonia, must be removed or rendered harmless by some means. Water exchange would have to be excessive to accomplish this, and ordinary

This is how a filter works. Water containing debris in suspension is sucked into the filter where it passes through a series of filtering devices designed to remove the debris. The water is then returned to the aquarium.

TABLE 3

Percent Free Ammonia at Different Temperatures and pH.

pH	Temp: 60°F	70°F	80°F
6.5	0.09	0.13	0.19
7.0	0.28	0.42	0.60
7.5	0.88	1.32	1.88
8.0	2.75	4.07	5.42
8.5	8.25	11.82	16.03

filtration, mechanical or chemical, can only cope with a limited amount. Hence the great importance of biological filtration, which without removal of anything converts toxic products into relatively harmless ones.

The course of the nitrogen cycle depends on a series of bacterial actions. Ammonia is attacked by a genus of bacteria called *Nitrosomonas*, that convert it into nitrites (salts of nitrous acid). These are in turn attacked by another bacterial genus, *Nitrobacter*, that converts the nitrites into nitrates (salts of nitric acid). Nitrates, being relatively harmless, may be allowed to accumulate, but they are in any case used by algae and thrown out with periodic water changes. Other bacteria can convert nitrites to free nitrogen, but this does not occur to a marked extent in the aquarium. There are filter systems that utilize such bacteria in order to eliminate even the production of nitrates, but this doesn't occur in the undergravel filter. The cycle is completed when nitrates are taken up by plants and some are eaten by animals, either of which end up, by elimination or death, as ammonia once more. The nitrogen cycle is aerobic, which means that it uses up oxygen, or rather, the bacteria concerned do so. They do not flourish in anaerobic or airless conditions,

hence a constant supply of oxygen is needed to keep them active, and they die eventually if without it.

The toxicity of ammonia has been the subject of many studies yet is poorly understood. Nevertheless, it is extremely poisonous to fishes, although they are themselves continually forming ammonia and excreting it, mainly across the gills. It is thought that

too much ammonia in the water slows down this excretion and causes a back-up of internally produced ammonia. Whatever the mechanism, gill necrosis (tissue death) and respiratory distress occur. Liver degeneration and susceptibility to disease follow even with doses low enough not to kill the fishes directly. A lethal level for different species varies, according to reports, from 0.07 ppm to 1.4 ppm. Any really measurable quantity is potentially dangerous, but the safe limit is usually quoted as 0.1 ppm measured as NH_3-nitrogen. The general run of marine invertebrates seem to be much less sensitive to ammonia; it is the fishes we must worry about.

The toxicity of nitrites is much better understood. They latch onto hemoglobin, the red oxygen-

THE NITROGEN CYCLE. The nitrogen cycle is the basic process for the recycling of animal waste products, not only in the aquarium, but in the whole world.

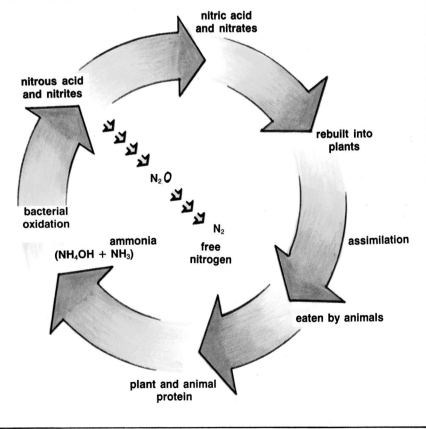

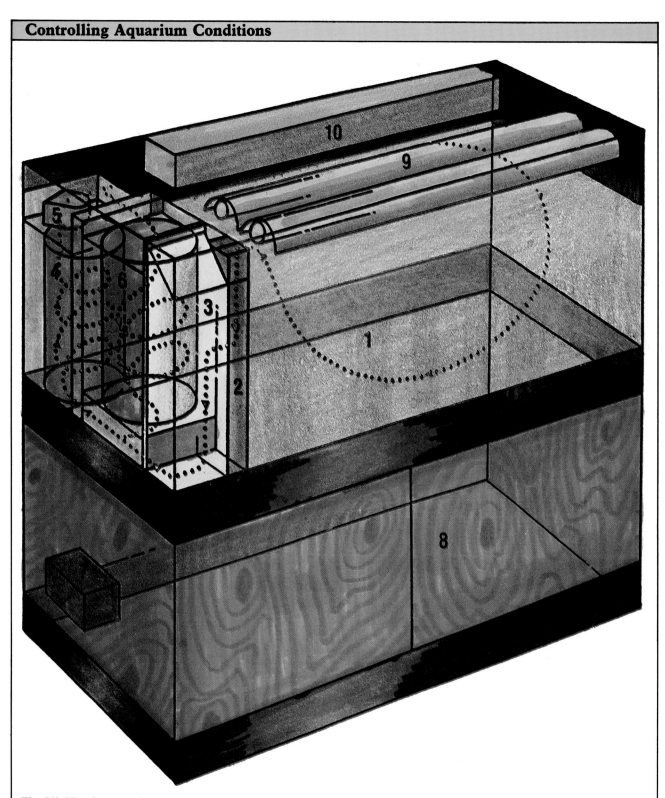

The MiniReef system features ease of maintenance and a beautiful, uncluttered aquarium. The main parts are as follows: 1. The Main Tank. 2. Heater and Carbon Chamber. 3. Protein Skimmer. 4. Aerobic Biological Spiral Filter. 5. Powerhead Pump. 6. Denitra Filter. 7. Air Pump. 8. Storage Cabinet and Aquarium Stand. 9. Lights. 10. Ballast Housing. MiniReef is a trademark of International Seaboard Corp.

carrier in red blood cells, and convert it to methemoglobin, which cannot carry oxygen, so the fish dies. Marine fishes are less sensitive to nitrites than are freshwater fishes, and some authorities doubt whether nitrites are much of a threat in the marine aquarium. The effect is dependent on salinity, not in the nature of the fish, since fingerling chinook salmon in fresh water are killed by 19 ppm; in salt water it takes 1070 ppm to do the job. However, fingerling chinooks are particularly insensitive to nitrites, rainbow trout are killed by 0.2 to 0.6 ppm, and so are smaller chinooks. What a pity we don't have data for reef fishes! To be safe, various figures of from 0.1 ppm to 0.25 ppm are quoted as allowable limits—this again is NO_2-nitrogen. In practice, if I can measure *any* nitrite I do something about it.

The toxicity of nitrates is very low—in fact, public aquaria that circulate the same sea water for even a year or two have reported levels as high as 300 ppm without obvious ill effects. For perhaps an overzealous degree of safety in the home aquarium, somewhere between 20 ppm and 40 ppm is usually regarded as the tolerable limit. Water changes usually take care of this, as does the growth of algae. No data seem to be available on the toxicity of either nitrites or nitrates to invertebrates, but the general feeling is that it must be low. For one thing, they do not typically have hemoglobin and their blood pigments, if any, are possibly not affected by nitrites.

Advanced Gadgetry

The marine aquarium has given rise to a number of pieces of equipment unnecessary in freshwater aquaria—and, many believe, equally unnecessary in marine ones! They are mostly of European origin and mostly connected with what is generally known as the "sterile" system, dependent on chemical types of control rather than biological ones. Let's look at the main techniques.

Protein skimmers, known also as air-strippers or foam fractionators, are, like biological filtration, adapted from sewage treatment practices. A voluminous stream of fine bubbles is passed through a column of tank water. If the water contains surface-active materials, organic compounds that form a film on the surface of the bubbles, these will produce a foam that can be skimmed off by a suitable arrangement. This foam will include many of the unwanted pollutants in the water; the more there are present, the more the water will froth up. Really polluted water forms a dark brown foam that may need daily removal. A really effective protein skimmer must be quite

deep, so that the bubbles rise through several feet of water, which means that it must stand outside the aquarium, from which water is siphoned to the base of the skimmer. Simpler versions can be placed inside the tank. Protein skimmers do not remove everything that charcoal takes out and have the advantage of leaving many trace elements behind. For this reason they may in particular be preferred in an invertebrate tank, usually for occasional use so as not to skim off too much plankton.

Ozone is O_3, an unstable form of oxygen (produced by an electrical discharge), that breaks down very readily and oxidizes anything ready to accept more oxygen, then becoming the usual oxygen molecule, O_2. It attacks bacteria, parasites, many pollutants, but also fishes and humans, and needs careful handling. It might well be expected to attack ammonia and nitrites, but it seems that this does not happen to a very useful

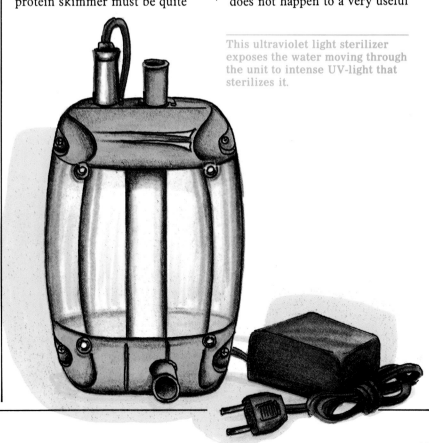

This ultraviolet light sterilizer exposes the water moving through the unit to intense UV-light that sterilizes it.

extent in the aquarium, despite claims to the contrary. It is hazardous to release ozone directly into the aquarium, and it should be metered into a protein skimmer or outside filter. It is introduced into the air supply by passing it through an ozonizer, a high tension electrical discharge tube that, in the more expensive instruments, can be adjusted to give a specific dose per hour. For additional safety, the return water can be passed through a carbon filter before entering the tank. Although ozone can sterilize the water that contacts it, it can only reduce the bacterial or other

content of the actual tank water and does not guarantee complete freedom from disease and parasites. If more than 0.4 mg per gallon per hour reaches fishes, it causes distress and destruction of the mucous layer and even of the skin itself; even lower doses kill biological filters. If you can smell it in the room, you are using too much. Ozone really belongs to the "sterile" tank setup. If you do use an ozonizer, put a dehumidifier between the air supply and the instrument, as moist air soon ruins the reactor. A long tube of absorbent crystals or anhydrous

flow of 10 gallons per minute over two 33″ slim-line tubes in series, with ⅛th inch water layer around them, is adequate for complete sterilization. However, do not imagine that a tank can be sterilized by such a setup, as there will always be masses of bacteria that never leave it to be treated, undergravel filter or not. The bactericidal range of UV is from 2000 to just below 3000 angstroms, with a peak at 2600 angstroms, so check that any equipment you purchase emits in this range. UV light can generate ozone, but only in the 1000-2000 angstrom range, which is therefore to be avoided as we don't want an unknown dose of ozone to be passed back to the aquarium. Tubes do not emit an effective dose of UV for very long and have to be renewed. The really helpful use of UV is in sterilizing water used for fish shipments, incoming water in public aquaria, and discard water from public aquaria and from dealers' premises, but not in the home aquarium.

I must confess that having tried all the above gadgetry, I now use none of it. That doesn't mean that it is useless; it is not. But it does mean that well-kept aquaria do not need such extra aids. They have their place in research and in special projects, like the breeding of some invertebrates and the keeping of "minireefs", which can be maintained by a variety of complicated procedures. For the ordinary fish and invertebrate keeper who wishes to maintain the animals and plants without necessarily growing corals or breeding bivalves, ozonization and UV sterilization are expensive and chancy. Protein skimming perhaps is worth a try, but it probably is less effective than simple activated carbon.

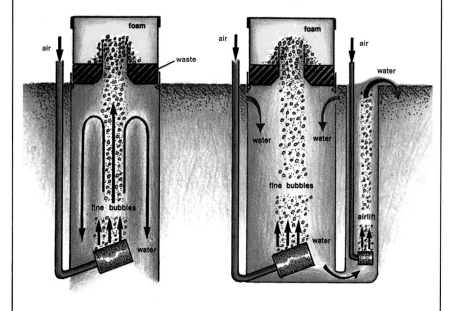

Protein skimmers are also known as foam fractionators or air-strippers. They work on the same basis as sewage disposal plants. A fine air lift of bubbles brings the unwanted waste to the surface where it coagulates and can be removed. The unit on the left is a simple unit; the compound unit on the right has a greater capacity for a larger tank.

calcium chloride suffices.

Ultraviolet (UV) irradiation is another sterilizing procedure. This must be applied outside the aquarium, as it would harm the inhabitants and can cause blindness in fishes and humans. It must be shielded from view, even outside the tank, but plastic absorbs it effectively. A thin film of water is made to run over long UV tubes so as to sterilize it; a

Biological filters can be anywhere—the base of the tank as in the usual undergravel filter, at the side or below the tank, or even above the tank, as in the popular European Tunze filter system. It all started with the undergravel filter that uses the substrate, sand or gravel, as a filter bed in freshwater aquaria. This essentially American invention was popularized for marine aquaria by Straughan and has become the standard method in the U.S.A. Not so in Europe, where a mixture of methods is normal and the use of the undergravel filter has never been very popular. Advanced European aquarists go to extremes to filter marine water in a series of treatment vessels so as to render it as pure and as well-oxygenated as feasible. This is because they are interested in culturing corals and other delicate invertebates and the concept of the "minireef" has so caught their fancy that you almost have to look to see which of the bank of tanks is the aquarium itself. This is not to decry such projects, because that is the way advances in aquarium

keeping are made and the fascination of our hobby is maintained. However, for the beginning and even many an advanced marine aquarist, the undergravel filter does all he finds necessary in a very unobtrusive and satisfactory way.

Undergravel Filters

In its simplest form, the undergravel filter consists of a perforated, raised plate covering the entire base of the aquarium and above which lies 2 or 3 inches of substrate, preferably not actually gravel in the marine tank, but crushed coral, crushed shells, or any suitable calcareous (calcium-containing) material. Silica sand, quartz, river sand, and gravel are not good materials to use because they do nothing for the pH and recalcification of the water. Dolomite contains magnesium as well as calcium and is preferred by some, but it is suspected of at least sometimes containing toxic material. The grain size is important, about $\frac{1}{10}''$ to $\frac{1}{5}''$ is best, with filter slots about $\frac{1}{20}''$, so that no substrate falls through. We want as small a grain size as feasible to

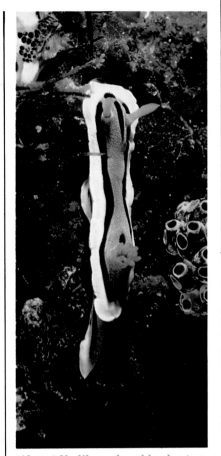

(Above) Nudibranchs add color to a reef tank, but most species are difficult to keep alive due to their dietary requirements. Photo by Cathy Church. (Below) One of the later developments in marine aquarium management is the biological wet/dry filter. These can be obtained in all well stocked pet shops. Photo courtesy of California Aquarium Supply Co.

offer as big a surface as possible for bacteria to populate, since the main function of the filter is to act as a massive culture of the bugs that transform ammonia via nitrites to nitrates. As this occurs the surface of the substrate granules will be so covered by bacteria that its actual nature will count less and less, but we might as well use calcareous material to get what benefit remains.

Water is removed from beneath the plate either by an airlift or airlifts or by a power head, which can give a greater flowrate. Airlifts are quite satisfactory as

long as they give a turnover rate in excess of 2 to 3 tank volumes per hour.

Commercially made undergravel filters are not always well designed. Some offer airlifts that are inefficient and clog up in salt water. Choose one that has a wide tube with an airstone inside the base, preferably served by an air-line running down the inside of the tube through a hole in the elbow bend at the top that directs the water across the surface. This arrangement allows easy access to the airstone. An uplift tube of ¾″ internal diameter is about right. When the setup is in action, adjust the airflow so that the right mixture of bubbles and water gives maximal flow—too much air and too little air both fail to do this. You can measure the flow by directing the water into a flask or bucket of known volume and timing it. Although a brisk flow is needed, both for the filter bed and for the inhabitants, most of which like a good water movement, too strong a flow is detrimental. This can easily happen with a strong action in a small tank and can exhaust the fishes.

If a power head is used that pumps the water from the filter without aeration, choose one that directs the flow briskly across the water surface, causing turbulence and not just falling down into the tank. This helps to aerate the otherwise rather exhausted water that has been depleted of much of its oxygen by the biological filter. In a tank served by such a motor instead of an airlift, it is advisable to include an airstone to help in further aeration and to remove the carbon dioxide layer that may accumulate over the water surface in a covered, unaerated aquarium. An air-cooled power head may assist in this as well, but a water-cooled underwater power head won't.

Heavy aeration from an elongated airstone rises briskly in this marine aquarium.

Reverse Flow Filters

The simple filter just described works extremely well as long as it is given any necessary periodic attention. Some aquarists feel, however, that it works the wrong way around. Instead of water passing down through the filter bed, taking perhaps some fine plankton with it that could otherwise feed invertebrates, it should pass up through the bed instead. Then, filtered water passes directly into the tank and pushes the plankton up into the water. They forget that this is deoxygenated water and that no surface stirring occurs except by other means. The biological action is just the same whichever way around the water flows, and the plankton is probably sucked up by the necessary uplift tube(s) and delivered to the filter from below instead of from above! Despite arguments to the contrary, there would seem to be no other differences in action, except that as the filter clogs with undissolved debris, which can happen, it will be all the more difficult to clean if the debris is concentrated at the bottom instead of at the top. It is fatal to clean a biological filter by removing all the gravel and washing it, because you then remove the beneficial bacteria. If clogging occurs toward the top, a stir-up and partial removal of

muck is easy, but not if it concentrates at the bottom. This muck removal can be done as you siphon off water when making a partial change. Just stir the gravel around with the siphon tube and suck up debris plus some gravel, later to be replaced.

Divided Tank Filtration

The big drawback to the usual biological filter is the ugly tube or tubes rising up through the tank that are rather difficult to disguise. In addition, if you wish to include carbon filtration to complete purification of the water, it has to be arranged separately and external to the tank if it is not to add another unwanted interior fixture. The same is true of any other addition such as protein-skimming and the necessary heater and airstones. All of these can be accommodated and the tubing eliminated by giving the aquarium a false back, a partition to be placed a few inches from the back glass in an otherwise conventional tank. If you construct or have made a special tank for this purpose, make it a little wider from back to front than usual so as not to lose too much exhibition space. Any tank larger than about 24″ x 12″ x 16″ (length x width x height) can easily accommodate such an arrangement, but 24″ x 16″ x 16″

would give the normal exhibition space, with a 4″ false rear compartment.

Instead of airlifts or water pumps, a slot at the base of the false back (about ½″ is enough) allows water from the undergravel filter, otherwise identical to the usual type, to flow into the rear compartment. In its very simplest form, this setup allows for heating and aeration in the rear compartment and for an airlift or airlifts, better still a submersible power head, to return the water to the front. If the false back is opaque (black or dark blue is best) nothing is seen from the front but a small return pipe or pipes sending the filtered water across the top of the tank. The false back should reach right up under the top covers so that nothing can get over it, and it should have any necessary slots to fit lightly around the return pipes. It need not be of very thick glass, as it will not bear much water pressure, having

water of approximately equal depth on both sides.

Further elaboration on this design is very easy, as the rear compartment can be divided into further segments. For instance, in, a 36″ x 18″ x 18″ tank with a 4″ rear section, this section can be divided into a 24″ x 4″ segment receiving the undergravel filter water via a 24″ x ½″ base slot, and two 6″ x 4″ segments. The first of these, next to the 24″ one, receives water from the latter since the dividing glass doesn't reach the top, being 1 or 2 inches below the level of the false back. This water then passes through a carbon filter (or any other filter) and in its turn runs into the second 6″ segment via another slot in the base of the glass dividing them. This glass partition reaches the top, but not the bottom. The filter material should sit on a small section of undergravel filter so as to raise it up and allow easy flow. In the last 6″ compartment sits the uplift or

power head returning the doubly filtered water to the front. Heating and aeration can best be accommodated in the largest compartment. Note that in addition to airstone aeration the water has to flow over the dividing glass so all of it passes near the surface and again is pumped onto the surface layer of the tank. The airstone aeration is as much to clear away surface carbon dioxide as to oxygenate the incoming water.

THE DRY-WET FILTER SYSTEM. Basically, water overflows into a mechanical filter containing floss or other filter medium. It then drips through a screen atop of which is a plastic grating similar to the kind used for fluorescent light dispersion. Inside this grating are small filter pebbles, stones, sand, gravel, or whatever else forms the basis for the biological filter system. The water after passing through this "dry" stage is gathered in the reservoir in the bottom and is further separated from heavy debris with a top and bottom baffle before it goes into still another cannister filter for further filtration and/or treatment.

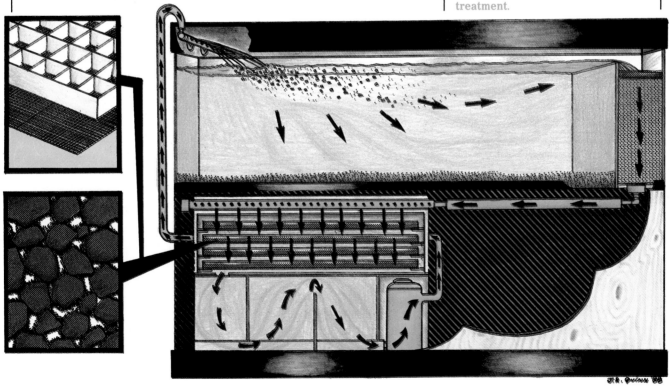

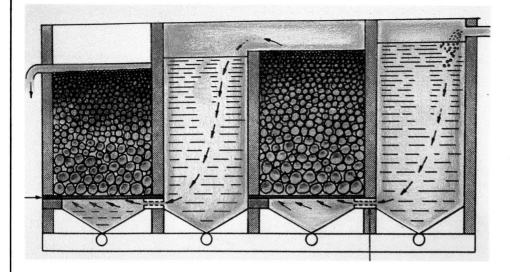

A four-chambered biological filter that serves as a flotation chamber to filter out biological debris. The water enters from the right and exits at left after having passed through two biological/mechanical filtering chambers in which nitrogen-fixing bacteria operate.

In a really large tank this arrangement can be doubled up. I have a 72″ aquarium with the same general design doubled up, so that there is a central compartment at back receiving water from the undergravel filter and a symmetrical filter and pump arrangement at each side of it. Heated and aerated water is thus delivered to the front at each top back corner, and an automatic switch enables both pumps to work together or to alternate so as to give a back and forth flow in the tank. The double arrangement is also insurance against breakdown.

Divided tank filtration offers all the advantages of an expensive custom-made setup of the type that usually has the "works" down below in the cabinet, without the drawbacks of much greater cost and of external tubing etc., that can always be a source of leaks or other troubles. Switches for the equipment, food, buckets, nets, and other accessories can be housed there instead. The system also has built-in checks on what is going on. If all is well, the water level in all compartments will be almost the same, with that in the pump or airlift compartment(s) perhaps a little lower than the

rest, as the carbon filter(s) must have at least a small positive pressure. If this level falls significantly it is time to renew the filter pad or pads, if not the carbon itself. If the level in the other back compartments falls, the biological filter is getting clogged and should be serviced. If in a newly setup tank the back compartments are at a lower level than the main tank, the return pump or airlifts are too powerful for the aquarium (an unlikely event, however). There seems to be an unexpected bonus in divided tank filtration in that in my own tanks, the oldest of which has been going for eight years and the newest for six, not one has ever needed a clean-out of the biological filter. It just doesn't clog up, perhaps because of the wide outlet to the rear compartment, but I have to confess that I'm not sure why myself!

The New Tank Syndrome

Biological filtration when first introduced to the marine aquarist can give him some horrible headaches. It exaggerates the functioning of the nitrogen cycle and, unless carefully managed, in the first few weeks the newly setup marine aquarium can cause

terrible grief. The tank is set up, everything is checked and seems in order, so the fishes are introduced and look happy in their new home. Calculations show that the tank is not overcrowded and all should go well. Not so! The fishes are

Undergravel filters may be purchased in sections and placed advantageously in various parts of the tank. Sometimes the gravel or sand may be too shallow in some areas for the proper functioning of an undergravel filter.

THE NEW TANK SYNDROME.
Unless you know a lot of water
chemistry, the beginning of a new
marine aquarium can be a
nightmare! It takes several weeks
for an aquarium to be able to
handle the biological filtration
system that converts the ammonia
to nitrites and finally to harmless
nitrates. Read the section on the
facing page dealing with "The
New Tank Syndrome."

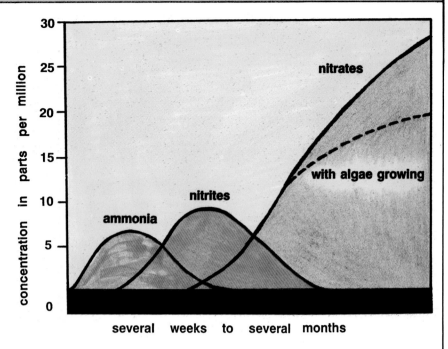

clearly unhappy after a week or
so, fins clamped, gasping,
scratching on decorations, and
most likely breaking out with
obvious disease. The result is
great anguish on the part of the
poor aquarist, who is lucky not to
lose many or all of his expensive
new fishes.

Why does all this happen? It is
because the new tank has an
insufficient bacterial population
to deal with the break-down
products of the fishes and their
diet, so a wave of ammonia
production occurs that may rise
to 5 or 10 ppm and be horribly
toxic. This wave will settle down
in the course of a few weeks,
depending on various factors, but
will be followed by a new one of
nitrites, less toxic but still
unwanted, that in its turn will
decline. All then will be well—as
long as there are survivors to
enjoy it! The biological filter has
had to undergo maturation, first
by the growth of millions of
Nitrosomonas bacteria to convert
the ammonia to nitrites, then by
the growth of *Nitrobacter* species

to convert the nitrites to
relatively harmless nitrates. That
is why we measure the nitrite
levels for the first few weeks in a
new tank, so as to follow the
completion of the maturation
process. If measures are not
taken to speed up the process, it
may in fact take two or three
months and be difficult to
endure.

The old method of coping with
the new tank syndrome was a
very gradual breaking in of the
tank. A handful of gravel from an
old tank or a sprinkling of garden
soil was added to the filter to
start bacterial growth as rapidly
as possible, and fishes were
introduced gradually over several
months until capacity was
reached. This was quite a strain
on the impatient aquarist, who
wanted to see his nice new
aquarium in full function.
Alternatively, some very tough
fishes or even turtles could be
added rather more rapidly to
speed the process—catfishes,
eels, some *Dascyllus* species, and
wrasses are examples. These had

to be replaced, if necessary, by
the desired species later.

Today, in place of old gravel or
earth we can use a culture of the
necessary bacteria to start things
off and then supply them with
the ammonia they need to
multiply, first the *Nitrosomonas*
and then the *Nitrobacter*. The
ammonia is supplied as
ammonium chloride, NH_4Cl or
ammonium sulphate, $(NH_4)_2SO_4$,
to be added daily until
maturation is complete. Then the
fishes can be added—all of them,
if sufficient ammonium salts have
been added. What is sufficient?
There are methods of calculating
this, but they are complicated. As
a sufficient rule of thumb, the
average aquarium needs the
following dosage.

Make up a 10% solution of
ammonium chloride or a 15%
solution of ammonium sulphate,
containing about the same
amounts of ammonia, and use
these to build up a concentration
of ammonia equal to that
expected from a full load of fishes
or other creatures. There is no

point in using less, or the introduction of fishes etc., will still have to be gradual. Start by

Species of the reef coral *Acropora* are sometimes included in reef tanks, but they do produce a lot of mucus-like material when they are "unhappy" which limits their use to the experts. Photo by Cathy Church.

adding 2 ml per day for every 25 gallons of aquarium water for the first two days, then add an extra 2 ml per day every two days until on day nine, 10 ml per day is reached and continue at that level until the nitrite peak is passed. This may take a month, which is about the quickest you can hope for. When the nitrite level is less than 1 ppm, stop the ammonia, give a partial change of water up to 50% to reduce the nitrate level, and put the fishes, etc., in within two days, or the bacteria will start to drop in numbers. The addition of 2 ml of NH_4Cl per day to 25 gallons of water is equivalent to 0.5 ppm of NH_4-nitrogen approx., so that by day nine we are adding 2.5 ppm per day. This is the turnover per day of NH_4-nitrogen that the filter is coping with by the time the ammonia peak subsides, subsequently an equivalent amount of NO_2-nitrogen. This in turn is around the expected output of a well-stocked aquarium, which will depend in part on the amount of food supplied and in part on the biomass—the weight of animals present.

Carrying Capacity

It is asserted that maturation of the biological filter by the ammonia method results in a greater nitrifying capacity, because by older methods a mixture of bacteria grows up from the start, some good, some useless, whereas when an ammonium salt is added the nitrifying bacteria are given a head start and crowd out the others. It is also asserted that a layer of ordinary filter floss under a thinner layer than usual of gravel results in a greater growth of bacteria because the filter floss offers a much bigger total surface for them to grow on. A combination of the two should, if these ideas are correct, result in a very effective filter. However, the filter floss wouldn't last forever, and what do you do then? I just don't know how correct either of these assertions is, but it's worth thinking about. The assertion about the filter floss is backed up by my own observation that an external filter that hasn't been

Upper photo at right and upper photo on facing page: Front and rear views of the same tank; the filtration system has a double layer spiral of synthetic under the rotating sprinkler.

Small crabs like this decorator crab can be droll inhabitants of a reef tank and provide many hours of enjoyment with their antics. Photo by Mike Mesgleski.

changed for a considerable period works well in a marine aquarium that doesn't have an undergravel filter, but one is again faced with problems of eventual renewal.

We can support far fewer fishes than in a similar freshwater tank. If we don't overfeed and if adequate maintenance is given, the maximum starting load for a conditioned tank is one 1″ to 2″ fish per 5 gallons of water, allowed to grow to 2″ to 4″ as time goes by. These limits can only be exceeded by a very experienced aquarist willing to give time to frequent water changes and general maintenance, but then not beyond the extent of one 2″ to 4″ fish per 3 gallons, not all to be 4″. The particular species also affects the issue—some tough fishes like lionfishes or anglers can be crowded more than chaetodons or angelfishes. Damsels and anemonefishes are good to start with; they are mostly quite tough and don't grow too quickly. All fish measurements are exclusive of the tail fin.

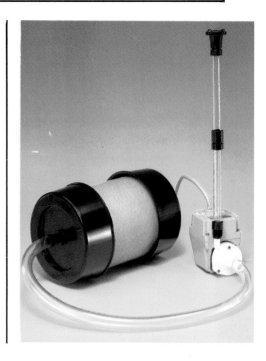

Large tanks, like the one shown below, can utilize garden pool filter such as the one shown to the right. The blue material is a foam sponge. The light blue pump pushes the water through the clear stem that aerates and moves the water very efficiently, as shown in the aquarium below.

Algal and Other Filters

Algae of any sort growing in a marine tank take up nitrates, and a good crop of them will keep the nitrate level well below the safety limit, particularly if they are culled frequently, thus removing the erstwhile pollutants from the aquarium. They also use carbon dioxide and produce oxygen when in good light, but like freshwater plants they produce carbon dioxide in poor light or in the dark. Luckily fishes are mostly inactive when algae are producing the CO_2, so little damage is done. Algae also release toxic substances like phenols from time to time, and in too intense illumination such as sunlight they can raise the pH over 8.4, which may cause distress if too much ammonia is around to be released in the free form. The best way therefore to use a heavy crop of algae is to remove it from the aquarium and grow it on pebbles or whatever in shallow trays close to fluorescent lamps. Then run the water passing over the algae through a carbon filter to get rid of unwanted substances before returning it to the tank. This is all a bit complicated and few find it worthwhile, but that's how its done.

European aquarists with their minireefs and other exotic projects have dreamed up all kinds of complex filters, of which the Tunze is the best known. It is suitable only for large aquaria and fits over the top of the tank. It is essentially a biological filter with extras, so designed that it can be serviced without disturbing the tank. It is not an algal filter and is shaded from direct light. It consists of a number of external filter beds in parallel, each acting as a biological filter. These can be cleaned and replaced one at a time, so that the rest are

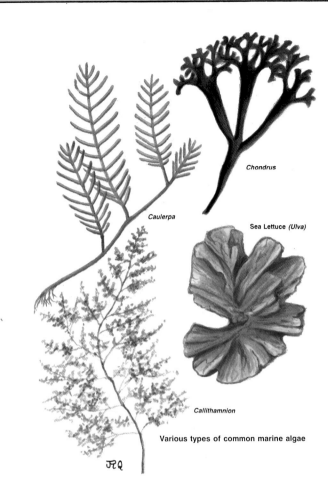

Caulerpa

Chondrus

Sea Lettuce (Ulva)

Callithamnion

Various types of common marine algae

Your aquarium shop should stock various marine algae and plants. Many algae are food for the fishes. Do not collect your own algae or plants as they may very well be contaminated with toxic substances or disease organisms.

functioning undisturbed. They supply purified water to a common return pipe that has a bleed-off by which part of the water is trickled through coarse gravel and is super-oxygenated before returning to the tank. Although it is unwieldy and expensive, it has enthusiastic support from many advanced aquarists. Some go further and cultivate bacteria that split off the nitrogen from nitrates or nitrites and ammonia and release it as a free gas. These work in anaerobic conditions, so a filter is rendered anaerobic and supplied with a source of carbon, needed to take up the oxygen or hydrogen released in denitrification, such as alcohol or lactic acid. The end result is water substantially free even of nitrates, but also of oxygen, so that it badly needs aeration before returning to the aquarium.

The dry-wet filter, now becoming known in the U.S.A., is an improvement on the undergravel filter in that the water is trickled in entirety through aerated beds of coral sand or plastic mat or beads, thus enhancing the power of the filter. It is then pumped over immersed gravel and back onto the surface of the tank. In some models, only the "dry", aerated part of the filter is included.

THE MINI-REEF: DO IT YOURSELF. One of America's greatest aquarists is Murray Wiener. He devised, tested, modified, and finally perfected this lovely mini-reef setup that has functioned perfectly for several years. Dr. Herbert R. Axelrod monitored the setup, photographed it, and provided the following captions.

The setup, shown in the topmost photo, is basically a normal marine aquarium plus a very shallow reef aquarium.

The top of the aquarium is covered with special fluorescent lamps that supply the necessary quality and quantity of light to keep the algae and plants thriving. The fact that the tank measures about one square meter (actually 36 x 36 inches) and is only 30 cm (12 in) deep makes the need for stronger lighting unnecessary.

Water action is generated by an inside pump that splashes the water onto both the front and the rear glass of the tank.

It is quite necessary for the pump to be very strong and be submerged. The quality of the aerated water is very important for the reef metabolism.

The tube outlet from the pump runs the entire width of the reef aquarium, thus guaranteeing that the aerated water is spread uniformly throughout the tank. The aerator pump in the back of the reef tank operates the same way.

This type of mini-reef is ideal for a location that benefits from a three or four sided view. By accumulating the living rock in the center of the aquarium, the view is excellent and the fishes really feel at home.

the tank and there is no danger of water spills. Hide the filter and tubes as well as you can and enjoy your aquarium. If you want greater elegance and can afford it, you can buy a custom-made job with all the works in the cabinet on which the tank sits, but it will cost you! Otherwise, get a suitable tank and fit it up as a divided one, or make or get one made. Either way, it will be much cheaper and free of external plumbing and the consequent danger of leaks or mishaps.

Whichever way you choose, you will have an outfit that should give you satisfaction, be easy to service, and will support fishes and most invertebrates indefinitely with no more than occasional water and carbon filter changes. It will probably not promote coral growth or be suitable for the attempted breeding of fishes, although they will spawn in it if suitably accommodated. For these, even greater water purity is desirable, but short of that the employment of other equipment is unnecessary. Encourage reasonable algal growth to help purify the water further, give a more natural appearance, and supplement the diet of the fishes and that of many invertebrates. Only disturb the biological filter when absolutely necessary and never uproot it more than partially.

Preferred Methods

For any but the advanced aquarist who wishes to experiment, there is no better choice than a combination of biological and carbon filtration. Together they purify the water very effectively without being difficult to manage. If you want to go to your dealer and purchase everything ready-made, buy a conventional undergravel filter and a separate activated carbon filter, which can be an internal one, so that everything is inside

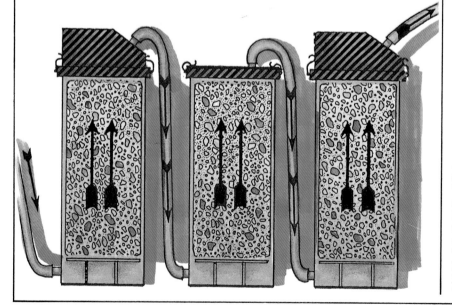

(Top left) Underneath the Murray Wiener mini-reef is a battery of cannister filters. Mr. Wiener has hooked them up using two cannisters with pumps and one cannister as a reservoir (center). The schematic to the left shows how the three cannisters are joined and the resulting flow of water. It is very effective.

Mr. Wiener sets up his marine tanks the same way as the mini-reefs. The living rocks are congregated in the center of the tanks, leaving room all around the rocks for life to begin. This gives four sides of viewing. It also increases the variety of ecological niches in which fishes and other living things can stake out their territories.

Hidden at the water level are the water pumps that create the water turbulence and surge that are characteristic of a coral reef. This strong water motion assists the filter-feeding invertebrates.

One of the most important steps forward in marine aquaria has been the introduction of all-glass tanks. Before that, did we have trouble! Putty used for aquaria, or aquarium cements as they used to be called, contained lead that leached into salt water and killed most creatures. Then came leadless putty, which was much better but still not ideal. The trick used to be to coat the seams internally with a protective substance such as bitumen, later with plastics. This tended to peel and allow seepage and was rarely satisfactory in the long run. The frames of such tanks also gave trouble unless made of first-quality stainless steel. At a later stage, some manufacturers dipped the whole tank in nylon after it was glazed. Sea water gets everywhere and attacks almost anything, so glass and the inert silicone rubber sealants now used are ideal, being two of the few substances safe to use.

The Tank

For the reasons just mentioned, we shall consider only all-glass aquaria, which can be purchased in a wide variety of shapes and sizes or can be made at home. I have never made a fish tank in my life, so I am not going to try to tell you how to do so. What I can tell you is that unless you know what you are doing it is just not worth making your own tank or, rather, trying to. Silicone rubber is tricky to handle, you only have a few minutes to do the job, and the cost of a well-made tank from your dealer is probably less than you will pay for the glass and sealer unless you use cheap, second-hand plate. The price of any but a very large aquarium is going to be one of the smallest items in a rather expensive hobby, so why worry? Even a divided tank won't set you back much more than usual, because the inner partitions need not be of thick glass.

The shape of a tank is always a compromise between elegance and biology. The greater the water surface, the better the gas exchange and the more creatures that can be housed, but flat "low" aquaria do not look as attractive as taller models. We must regard the extra water in a "high" aquarium as so much dead material, giving no additional biological advantage or fish capacity. The usual compromise is to have the height a few inches greater than the width, say 36″ x 16″ x 20″ rather than 36″ x 18″ x 18″, the double-cube of older aquarists. Really large tanks are usually longer in comparison with their width and height—i.e., 60″ x 18″ x 22″, not an unwieldy 60″ x 30″ x 30″ or anything like it. Deep tanks need very thick glass and are difficult to service; wide tanks stick out into the room too far and are also difficult to service unless you can get all around them.

Small tanks need more

The beginning of a marine aquarium is, of course, the aquarium itself. Only an all-glass aquarium can be recommended for salt water. The tank should be labelled with the name of the manufacturer and should have a guarantee. The glass should be appropriate for the height since the other dimensions of the tank have little effect upon the need for strength.

Some fishes are nocturnal, that is, they come out primarily during the night to feed. Such is the case with most squirrelfishes. Shown here is the Longspined Squirrelfish, *Holocentrus rufus*. Photo by Mike Mesgleski.

attention than large ones and are more subject to fluctuations in conditions, including temperature unless in an air-conditioned room, even if heated. In hot weather, a large tank takes a long time to warm up beyond the desired temperature and is much safer than a small one for that reason; and when the temperature drops it won't rush down and chill the inhabitants. Large tanks do not pollute as readily, suffer less in a power failure, and look better than small ones because both need a minimum of 2″—3″ of gravel at the bottom for adequate biological filtration. On the other hand, small tanks can be supported on most pieces of furniture or a shelf, whereas large ones can weigh a great deal. Small tanks can have considerable water changes without breaking the bank and thus to some extent make up for their greater "pollutability".

I recommend a tank of not less than about 25 gallons but, for a beginner, not more than about 50 gallons. This is because of cost more than any other factor, of both the equipment and the fishes or invertebrates, and the likely loss of some of the creatures in amateur hands. Marine life costs more and more to purchase, and a big tank takes some populating if it is not to look pretty meager.

To calculate the gallon capacity of a rectangular tank, multiply height by width and length and divide by 231. For the 36″ x 16″ x 20″ tank mentioned, we have just under 50 U.S. gallons, but this is supposing only water in it, whereas some water volume will be lost by the presence of gravel, coral, and equipment and the fact that the tank will not be filled to the brim. Usually this will lower the actual water volume by 20% in the size of tank considered and by more in smaller tanks.

Well-made Tanks

First of all, a good tank is made of thick enough glass, and the main determinant of the necessary glass thickness is the depth of water it has to hold. Area means nothing; 1/4″-plate glass will hold back a lake if it isn't more than 18″ deep where the glass is and if the glass is properly supported. Up to a foot in depth and 30″ long, thick window glass is sufficient (about 3/16″ thick), with a slightly thicker base, say 1/4″-plate. From 12″ to 18″ in depth 3/8″-plate sides and bottom are safe, but tanks less than 15″ deep can be made of 1/4″-plate if unscratched. Above 18″, 1/2″-plate is safe, but new, unmarked 3/8″-plate can be used to 24″, (not salvage glass, however). Never accept deeply scratched salvage glass in any tank, as the intact skin of the glass gives it its greatest strength. For the same reason, guard against damaging the glass yourself when the tank is in use.

Large aquaria should have the inside seams lined with thin glass rods that form a seal with the silicon rubber and help to prevent leaks. They must also have supports across the top, one in the middle about 3″ wide for a 36″ tank, and two, each a third of the way along for a tank in excess of 60″, and 4″ to 6″ wide at least.

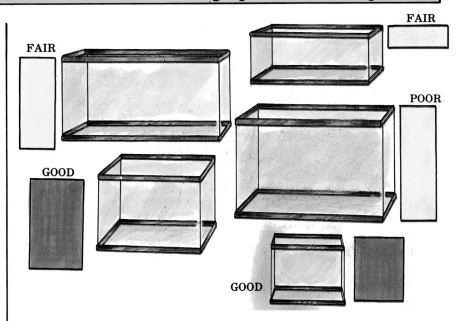

Surface area is an important consideration. The above tanks are used as examples with the orange and yellow blocks representing the surface area. Clockwise starting with the uppermost: this tank has low sides but is too narrow—it is a fair choice; an appealingly high tank with a width that is entirely too narrow for its gallon capacity—a poor choice; a square tank offering a good surface area—a good choice; although more rectangular than the preceding, this tank offers good surface and is also a fine choice; a nice looking long but still too narrow tank—a fair choice.

These prevent bowing of the long sides and also enable top covers to rest conveniently. They should be countersunk about 1/2″ below the top edge of the glass. Stout top cover glasses must cover the whole of the top except for small triangles cut at back corners to allow filters and leads to enter. At least one cover should have a handle cemented onto it for convenience in feeding.

The top covers guard the lights, prevent fishes from jumping out, and also prevent splash and excessive evaporation. They equally importantly prevent things from falling in, such as sprays and cigarette smoke, even cocktails. They should be countersunk as are any cross supports, and are best arranged by cementing a 1/2″ rim of glass to the inner surface of the ends or sides at such a depth that the top of the covers is 1/2″ below the top of the sides, with similar rims on the undersurface of cross supports. This arrangement saves a lot of bother—splashed or evaporated water drips back instead of running over the sides, the covers cannot slip off, and you can clean the outside of the covers with a little fresh water and any run-off goes to help make up for evaporative loss from the aquarium. As the tank ages, the glass tends more and more to acquire salt deposits on top covers as splash evaporates on the inner surface and in part creeps up between joints. I am very lazy, and I get rid of much of this by just pouring some fresh or distilled water over the whole top, which dissolves much of the deposits as it runs in. Eventually limey deposits that are hard to shift develop, but they can be dissolved easily in a little dilute hydrochloric acid—not over the tank though!

Take a careful look at any aquarium before buying it to see that it conforms to the above

Most well-equipped aquarium shops have a variety of decorator tanks that are esthetically pleasing as well as useful. Most of the unsightly equipment can be hidden away in included cabinet space. Photo courtesy of California Aquarium Supply Co.

living room floor. Bookshelves take quite a load, and an aquarium that will fit onto one is going to be O.K., but beware of flimsier shelving.

Whether your all-glass aquarium will sit on a special stand or a piece of furniture, the first requirement is that the support must be absolutely level. Use a spirit level to ensure that this is so, from end to end and back to front. Even a very slight tilt, although insufficient to stress the tank significantly, can look dreadful when the tank is filled and the water is not level with the glass. Although framed tanks can sit on runners and be unsupported elsewhere, all-glass tanks must sit on a completely flat surface. Between the tank and the stand place ¼" to ½" styrofoam or similar waterproof compressible material so as to take up any small irregularities that may exist. Tape the styrofoam in position temporarily so that it doesn't slip when you are placing the tank over it, then with any but a very small tank get a friend to help you lift it on. Hold the tank at the base and see how good you are at getting fingers out of the way as it is lowered. Make sure that the tank rests everywhere on the styrofoam sheet or sheets, which are best cut a little bigger than the tank base, as they can be trimmed later if necessary. I find it best to have a stand or cabinet with a beaded edge that will cover the front and sides for about 1" from the base, so as to hide the styrofoam and the bottom edges of the tank. Keep the back clear and the tank can be more easily placed in position.

Look very carefully once more before you put the aquarium on its stand, to detect any previously missed scratches and see that the best face is in front if there is a choice. Some tanks have a

requirements—particularly any large tank. Make sure that it is absolutely square—not in shape, but in having strictly right-angled joints. It doesn't matter whether the sides rest on the bottom or the bottom sits inside them; what matters is that everything must be carefully supported however well made, as we shall see later. See also that all edges have been honed and will not cut, that the sealant is everywhere intact, with no bubbles or unevenness anywhere and no evidence of joints or channels through it.

Placing the Aquarium

Water weighs about 8¼ lbs per U.S. gallon, coral, rocks, and gravel even more. Glass is heavy too, so that an aquarium of any size is very heavy, possibly too heavy for the shelf, cabinet, or even the floor underneath it. With average construction, no problem with the floor is likely to arise with tanks up to 4 ft in length, and even longer ones will give no trouble as long as they are not in excess of 18" × 18" in

width and depth. It is safest to have the aquarium against a wall, where the floor is strongest, and over stout beams if they exist. Take this advice in the case of a large tank so as not to emulate the unlucky guy who returns to find the basement full of debris and dead fishes and a hole in the

When placing your aquarium on the stand, be sure that the legs of the stand are located on a very stable part of the floor, preferably one that is supported underneath. Then a piece of styrofoam should be laid on top of the stand and secured with tape so it doesn't move when the aquarium is eventually placed on it.

predetermined front and back, as with a divided tank, but with others there is usually a choice. If necessary, clean the inside of the aquarium thoroughly with lukewarm water, using generous amounts that can be siphoned off. Then pour in some more water to wash away the first, possibly polluted water.

Setting Up

Proceed in the following order to make your work easy and as rapid as feasible:

1. Place the undergravel filter in position, making sure that it fits snugly against side glasses and the back, but preferably leave a half-inch or so margin at the front so that when you add the "gravel"—really coral sand or shell-grit, etc.—it will hide the filter from view. Connect any vertical airlifts or similar tubes that serve the filter. Now cover the filter with a double thickness of plastic flyscreen, tucking the edges neatly downward and well over the possibly open front where it doesn't meet the glass. However, it is better to have as the front that part of the filter plate that may not have had to be cut to fit and so is impervious. If you don't mind seeing the edge in front or will camouflage it in some way, it is neater to have a filter that exactly fits the tank you have chosen. If you make up your own filter, it will have open edges to be dealt with all around.

In a divided tank there are no worries about uplifts. All that has to be done is to see that the plate fits snugly and is not obstructed at the entry to the back compartment.

2. Place well-washed "gravel" gently over the filter to form the filter bed, sloping it a little from back and sides to the front so that it forms a shallow half-basin several inches deep at the back and sides and not less than 2″ to 3″ deep in the middle front. This looks pleasant and assists the mulm to collect towards a convenient spot to be siphoned off. It also helps the filter to collect more evenly from the base of the tank, as it has a longer "pull" from the front where the substrate is thinnest. Of course, depending on the fishes, etc., you keep, this arrangement may not stay put for long, but do your best! Coral sand and suchlike under water do not weigh much and are liable to be rearranged by

such fishes as wrasses and gobies, even by damsels or anemonefishes that often feel like digging around.

3. Place rocks, coral, and other decorations where you have decided they should go. It pays to make a little chart of what you are going to do beforehand so as to achieve a desired result. In an ordinary tank, budget for heater, thermostat if separate, air-lines and airstones, and an internal carbon filter if it is to be used, all to be placed in as inconspicuous a position as possible behind rocks or coral. Remember to keep the heater free of contact with glass and with a good water jacket around it, or it may overheat and fuse. If you lay it on the bottom, keep it above the actual gravel level with small pieces of coral or some other device. Nobody seems to manufacture a cradle for this purpose. Be suspicious of the suckers often supplied with heaters to hold them in position;

When setting up the tank, placement of decorations is of course partly dependent on individual taste, but practical considerations must also be taken into account.

they may lose their suction a bit too readily. Weigh down leads and air-lines to stop them from floating up when you fill the tank.

In a divided tank you don't have to worry about placement of equipment at this stage; it is all going into the back. Before filling the tank it is desirable to put the equipment in position with the same precautions just outlined. The only exception is an air-line for a front airstone if you so desire, but an airstone or two in the back will best aerate the filtered water. Otherwise, the heater and airstones go into the center back compartment, a carbon filter into the filter

compartment if you have one, and the return airlift or pump into or over the final compartment for the return of water to the front.

4. In this first filling of the aquarium, it is not necessary to dissolve the salt mix you will probably be using prior to filling the tank. Make a rough guess at the actual gallon capacity allowing for substrate, rocks, etc. (about 20% off the theoretical is a good guide), and dump the salt into the tank at 4 ounces per gallon or somewhat less if you wish to use more dilute water. Later a check will show how good a guess you made; adjustment is very easy one way

or the other by using a hydrometer. The only drawback to this lazy method is that the salt may for a few minutes clog the filter, which won't be working very well and cause some alarm— don't worry, it will soon correct itself. Expect some of the calcium salts to take time to dissolve. You have all the time in the world, as you won't be putting anything alive into the tank for days.

Fill the aquarium very gently by running a hose into a basin or low container placed in center front; as it overflows from the container the water will not disturb things too much. It is best to run the hose for a few minutes beforehand to get rid of any standing water in the pipes that may have picked up metallic contaminants. Be wary of copper piping, and never use water from the hot water system to hasten dissolving the salt. Old copper piping may be safe, but if you contemplate keeping invertebrates test it to be sure. Remember also that the water may need treating for chloramine if that is a local problem. Just follow the recommendations already given. Don't worry about simple chlorine; that will soon blow off as you start things running. For super safety, only half-fill the tank the first time, just in case there is a leak (most unlikely but still possible) and wait a few hours to complete the filling.

If you use natural sea water, pour it into a receptacle as above using a spouted vessel of some kind if possible. A plastic watering can is excellent. In a divided tank the water, fresh or salt, can be poured into the back compartment, where it will flow by reverse flow upward through the filter and give no trouble with clogging by salt and no worry about spoiling arrangements. The divided tank has a lot going for it!

5. The filter by now needs seeding with the right bacteria. These are best purchased from your dealer or supplied in the form of a handful of gravel from an old *disease-free* tank. Much best the former, as how sure can you be? I would rather put some soil from the garden in (just a few teaspoons are enough) or some gravel from a freshwater tank, as the same bugs are present, even though they have to adjust to the new conditions.

Leave adding any ammonia for the moment, as we want to see that everything is working, so switch it all on and see. See that the filters are going well and the airstones are giving the right-looking stream of small bubbles, not too brisk, just a nice, full stream. Check that the temperature rises to about 76°-80°F; anywhere within that range is O.K. When the temperature is right, test the density of the water with a hydrometer, which should read about 1.022 or 1.023 if you want to copy normal sea water, not less than 1.018 if you decide to have the water more dilute, or there will be trouble with introducing new fishes and with many invertebrates.

Maturation

As noted already, you *can* mature a tank by the old methods, a few hardy fishes or a chunk of dead meat or fish left in for a week or two (rather smelly), but why do it? The ammonium chloride or sulphate method is simpler, quicker, and allegedly results in a greater fish capacity. There is no point in trying to test for ammonia as you are putting plenty in and will surely get a high reading for a week or two. Anyway, test methods are poor. Start checking the nitrite level after two weeks; it should be showing up in the 10-15 ppm range now or very soon. Go on

checking nitrite every other day until it falls to the safe level of about 1 ppm around three to four weeks from the start. Don't worry if it stays high for longer, as all kinds of factors influence the time taken, such as the temperature, amount of seeding, speed of filtration, and nature of the substrate.

When the nitrite has fallen, check the pH, which should be between 8.0 and 8.3; it if it isn't, adjust with sodium bicarbonate. Stop adding ammonium salts, check the pH again the next day, and if all is well add the fishes— all or nearly all that you intend to start with. If you can't get as many as you would like to have eventually, because of cost or non-availability, don't worry, it doesn't matter. You will have to add the subsequent ones not more than a few at a time, only one or two in a small tank, three or four in a really large one. This follows because the filter bed bacteria decline if not fully required and will only build up

again rather slowly. Add the new batches at not more than weekly intervals. The advantage of adding all the fishes at once is that they will not have selected territories and be prepared to guard them and will be much more likely to settle down peacefully. A new fish introduced to an existing population often has a hard time and may even be bullied to death. Remember the expected carrying capacity of your tank when starting up—i.e., not more than one 1″ to 2″ fish per 5 gallons of water. If you wish to start with bigger fishes (I wouldn't) cut down the number considerably; a 4″ fish needs 10-15 gallons. It is not possible to give precise-sounding rules as for freshwater fishes, as so much depends on the fish species and on the particular aquarium.

It is inadvisable to populate a new aquarium with living corals, anemones, and other rather touchy invertebrates, but you can introduce most crustaceans and starfishes if you wish. Wait a

Once the tank is ready for occupancy, it should be tested with sensitive invertebrates like living coral, a sea anemone, and some higher algae like *Caulerpa*. If these thrive for a week or so, you can assume that the tank is safe for the living rocks, fishes, etc.

month or two before putting the others in, and then be frugal about it. Exactly why this is so baffles me, but experience shows it to be the case. Some further conditioning factors must clearly be involved, but nobody has explained what they are. Oddly, you can start a "natural" system aquarium with masses of living rock, even using a salt mix and not natural seawater, but even then larger creatures should be added slowly as in the unconditioned filtered tank, with fishes last of all. We have a lot to learn.

Periodic Maintenance

The frequency with which an aquarium needs attention, other than feeding and a general check

that all is well, which should be made daily, depends on several factors such as the illumination, number of creatures, light or heavy feeding, and filter efficiency. The following are therefore only rough guides to running a tank.

Weekly Clean glass covers and, if necessary, the inside glass of the tank. Algae can be removed with a sponge or ball of cloth on a stick, or if difficult to get rid of, by a razor blade scraper. The metal will not contaminate the water as long as a new blade or a very carefully preserved used one is employed each time. After the first month, add 1 level teaspoon per 25 gallons of sodium bicarbonate to maintain the buffer capacity.

Monthly Siphon off 20 to 25% of the water in tanks under 40 gallons or 10 to 15% in larger tanks, depending on the animal load. Take the opportunity to lift rocks, etc., when possible to siphon any debris from under them and stir up the gravel a bit. A convenient siphon is a glass or rigid plastic tube about ½" in diameter, about as long as the depth of the tank, with a rubber tube attached that is long enough to reach the floor. A short piece of rubber tubing on the tip of the siphon prevents damage to it or to the aquarium contents. A rubber tube is much easier to manage than a plastic one and can be pinched to control the flow. Check the state of the carbon filter and renew the filter pads or flow if necessary. It is not necessary to renew the carbon unless you are using very little or it is not first-grade. Check the pH, nitrate level and specific gravity.

Quarterly Completely renew the carbon filter, first washing the new carbon in fresh or salt water. If the coral or other decorations are overloaded with algae, either replace part of them with bleached coral or rock or take some out and scrub off much of the algae. Some algae on coral and rocks looks nice, but too much looks very dull. This procedure may need to be undertaken more often than quarterly. Bleach in the sun if possible; if not, use a household bleach and rinse several times very thoroughly before returning the coral to the tank.

Up to a year or more Siphon off about one-third of the gravel if it seems necessary, rinse it, and return it to the tank. Indications to do this are much detritus swirling up into the water on disturbing the gravel and

The classic aquarium layout also works for the marine tank. In order to save investment, the rocks can be placed in a horseshoe shape covering the back and sides of the aquarium. The view of the tank can only be appreciated from the front. Living rocks are not necessary, as the diagram below indicates.

Slate		Plastic plants	
Rocks		Dried coral	
Printed background		Artifical plastic coral	

Tank layout viewed from the front

blockage preventing adequate water flow—gurgling airlifts indicate this, for example. After doing such a thorough siphoning, feed lightly for the next week in order to reduce the load on the undergravel filter, as it cannot cope with as much pollution as it could previously until the bacterial population has built up again. It is best to do the siphoning in several stages, leaving say two-thirds of the surface undisturbed each time. This is because the top layers of the gravel do most of the work, and if all of the surface is removed at once it removes a lot of filter capacity. If feasible, space out the siphonings a week apart.

This nudibranch, probably a species of *Casella*, wends its way over some of the other animals in the reef tank. They often will lay eggs on the glass sides of an aquarium where, if they are not eaten, can be observed as they develop.

Magnificent ceramic fishes are being produced in China and other areas of the world. They usually are mounted on a piece of treated, dried coral and can be put into an aquarium as a realistic decoration. The fishes usually ignore such decorations.

You will rarely discover the history of the fishes you buy from a dealer. In fact, he very often will be as ignorant as you are if he purchases them from a wholesaler. Yet, as we become more conscious of the damage done to the environment and the wastage involved in the collection of many marine fishes, we should make every effort to influence the trade to set its house in order as far as this is possible. This is also in our own interest, as fishes collected with drugs or coming from an exporter who handles them in a careless fashion and loses many before even shipping them are not going to be in good condition. Strangely, it is also possible for an exporter to treat his captures too kindly and not condition them to the journey they have to make and the circumstances in which they will be kept. I know of one dealer who used to keep his fishes in a series of enclosures in communication with the sea, with the result that they could not stand the sudden change to shipment and aquarium conditions and frequently died.

There is no relationship more important than the one between yourself and the dealer from whom you purchased your original setup. The dealer should have many large marine tanks of his own. These tanks should be healthy and active. The fishes, plants, and invertebrate animals should be clearly labelled with their scientific and common names, as well as the price. The dealer should be capable and willing to help you.

Many fishes change thier color pattern as they mature. An informed dealer can tell you (or show you) what the adult will look like even if he has to find the picture in a book. You will want to know how large a fish will eventually be and how to feed it. You should be interested in what it will kill and what will kill it!

DRUGS

The worst thing that can happen to a fish is to be collected with the aid of cyanide, and it is also the worst thing short of dynamiting that can happen to the area in which he is caught. Cyanide is a poison that has the effect of paralyzing fishes in doses that are not immediately fatal and so making them easy to catch. It also kills off other creatures around the capture site. Some of the fishes caught with cyanide die rapidly, some live a while and die later on, some lucky ones survive relatively unhurt—it all depends on the dose they happen to receive. Survivors are often so poisoned that they cannot digest their food; they can look normal, eat at first and therefore have full stomachs, but eventually starve.

So with a batch of cyanide-caught fishes, the unfortunate purchaser often cannot at first tell that anything is wrong, but many will probably die within a few weeks. Only a refusal by dealers to purchase fishes not guaranteed to have been caught without cyanide, or preferably any drug, can help. You can help by refusing to purchase fishes not under such a guarantee, but regrettably this will restrict your choice of purchases rather severely at present.

Shipment of Fishes

Although it is possible to catch a fish and put it straight into a good aquarium and see it flourish from there on, fishes that are going to undergo holding, often a long journey, and then holding again, perhaps in several hands,

before arriving into your tank need special treatment. A newly caught fish put into an aquarium and fed its natural food or something acceptably like it will go on eating. Put into a holding tank in crowded and unnatural surroundings it starves for a variable period. If it is to be shipped quite soon, the exporter will not even try to feed it, as it then doesn't foul the transport water so much. If he is likely to keep it for some time he will have to teach it to eat whatever he offers, or lose it. Fishes can go for a long time without food, and an unscrupulous exporter will take advantage of this and not worry about their chances of recovery once he is rid of them.

Fishes for export, properly handled, will have become accustomed to feeding in captivity. Before shipment they will be starved for a few days to reduce pollution during shipment, placed into individual plastic bags in the case of any but tough small specimens that won't harm each other, probably tranquilized, and given oxygen. They are then shipped in insulated containers for reception by the wholesaler or individual dealer. During a journey of any length the shipment water becomes charged with ammonia, but this is usually rendered fairly harmless because the pH of the water falls as CO_2 accumulates. On arrival, the fishes should be gradually introduced to the tank water in which they will be held so as to adjust the pH without shock and dilute the ammonia sufficiently to render it harmless even at the higher pH of a normal tank. The fishes have been stressed, but they should recover within a few hours or a day or so and rapidly learn to eat again.

Selecting Your Fishes

When buying a fish the first thing to do is to look at the whole tank and if anything seems wrong with any of the inhabitants, don't buy from it, even if the particular fish you want seems perfect. It may be infected with whatever is affecting the other fish or fishes. Ignore the odd torn fin or tail, as long as it is not inflamed, as bickering among marine fishes is very common and not to be worried about unduly. The tears will heal quickly. If a piece is actually torn from a fin or tail it is a different matter, because the color may not return when the appendage heals, leaving a clear spot or margin that spoils the appearance of the fish.

Large nets trap coral fishes that are chased into them and then small nets capture the fish for export. The Philippine Islands have been condemned because some unscrupulous dealers use cyanide to daze the fish during the capture process.

If the tank from which you select your fish looks in order, inspect your proposed purchase very carefully. It should be well-fleshed, with no sunken abdomen or wasted musculature around the head or back, giving the so-called razor-back appearance indicative of a long period of starvation. Cyanide-caught fishes nearing the end may show a full stomach but a razor-back, because their stomachs may be full of undigested food. If a fish looks a bit thin but is not in any advanced state of wasting, ask the dealer to feed it and see if it eats. If it does and you are very keen to have it, you might risk the purchase as it will probably recover. It is unwise to buy any fish that does not eat, even if it looks well, so it is a good idea to ask for any fish you propose buying to be fed and to watch the result.

Look critically at the body and fins of the fish. There should be no blemishes, torn scales, or signs of infection. Try to look obliquely at the fins and body, particularly for *Oodinium* (velvet disease) and *Cryptocaryon* (white spot). *Oodinium* especially is hard to see sometimes and shows up best if you can get the fish facing toward you, when it appears as tiny whitish spots or a rough whitish coating, hence the name "velvet". White spot looks bigger, up to small pin-head size, and is easier to detect. Look also for any knobs or discolored areas that might be signs of other types of infection. A good dealer will have quarantined his fishes and will probably have treated them routinely with medication against such diseases, but he may not always have been successful. Many only treat for a disease if they spot it and may have missed

Snug-Fitting
Styro Lid

Plastic bag filled with oxygen and
medicated with M-Green.

Inner Styro
Container

Outer Carton

TYPICAL FISH SHIPMENT BOX

(Above) Once the fishes are collected and enough are accumulated to make a shipment, the fishes are placed into heavy plastic bags about a quarter filled with water. The balance of the bag is filled with pure oxygen. The whole bag fits very snugly into a molded styrofoam box with a suitable cover. The styrofoam box itself fits into a heavy corrugated box that has precautionary markings that the box contains live tropical fish. (Below) Collecting marine aquarium fishes is a big series of small businesses. Usually teams of 2 to 12 men go onto the reef with a small boat and Scuba gear. The fishes are collected and brought to shore where they are sorted and accumulated until enough are available to make an economical shipment by air. In many areas of the world wholesalers buy fishes from the collectors and solicit business world-wide. *Tropical Fish Hobbyist* magazine contains advertisements from such dealers.

As the market for living rocks grows and the areas in which they may be collected shrinks, a new business is developing for the farming of living rocks in non-reef areas. Tropical seas are laced with dead coral that is planted in an area that is rich in invertebrates. Soon the coral is covered with growth and sold as living rock.

something. Some keep all fishes in antibiotic- or copper-treated tanks that may suppress diseases but not cure them; they are the hardest to cope with effectively.

The head of a fish often tells a great deal. The eyes should be clear and should not protrude; a misty appearance means infection or toxic conditions, and you want neither. Make sure that you see both eyes, as one may be quite normal and the other not. The mouth and gills should be free of

any signs of infection or damage and should show only moderate respiratory movements, not gasping or respiring rapidly, with the gills moving gently and not opening unduly wide. The normal respiratory rate in the aquarium is up to 100 per minute, anything higher is suspect and may indicate gill infection or damage. Parasites often find the gills first and are impossible to see in the living fish unless they actually protrude.

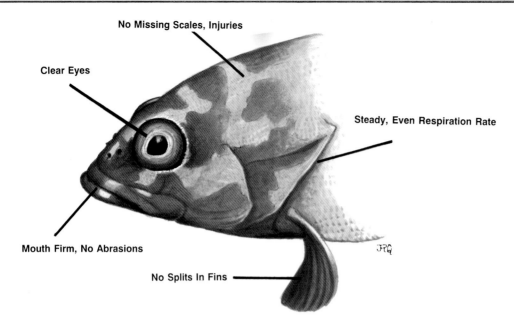

No Missing Scales, Injuries

Clear Eyes

Steady, Even Respiration Rate

Mouth Firm, No Abrasions

No Splits In Fins

(Above) You should learn how a healthy marine fish looks. Groupers, like the one shown here, are among the most hardy fish. The characteristics for you to observe are labelled. (Below) Here is a wholesaler checking in a shipment. Laborers open all the cartons but the fishes are not touched until the expert arrives. The expert quickly examines each bag and assigns it to a separate tank. In this way diseased and antagonistic fishes are contained to a minimum. Diseased fishes usually are destroyed as wholesalers cannot afford the risk, space, or time to treat sick fish.

Fish behavior is very important. A healthy fish is alert, with fins that are neither clamped nor permanently erect. The fins should be in motion frequently as the fish moves around. Of course, some fishes are naturally secretive or nocturnal and hide away, but not the typical reef fishes that you are likely to be purchasing, except for some wrasses, mandarins, blennies, and gobies. Other fishes should be swimming up in the water, frolicking around and not lurking in corners or hiding away somewhere. They should be hard to catch, but beware of the assistant who is not skilled enough to do so without undue chasing and possible damage to the specimen he is after. I don't know why, but many a dealer seems to employ a single net about the size of a soup spoon and chases a fish all over the tank endlessly before catching it. Two

These are two fish of the same species. Can you differentiate the sick one from the healthy one? The fish to the right is healthy. It has upright fins and swims actively (an active fish can't swim with clamped fins). The fish to the left has drooping fins, a sagging body, and is listless in the aquarium.

nets of reasonable size make the job much easier and stress the fish far less. Incidentally, if you are buying a small fish from a tankful of the same species don't expect the dealer to chase a particular fish for you, but do expect him to let you see the one he has caught so that you can approve of it.

Selecting Species

Marine fishes are belligerent. Don't be misled by seeing them crowded together in the shop; they have had no opportunity to select the territories typical of reef fishes and to develop the habit of guarding them. Some fishes live in communities, such as many anemonefishes and *Dascyllus* species, and so can be expected to tolerate one another even when settled into a decorated aquarium, others are highly territorial and drive off others of the same species, although they may tolerate fishes of a different species. Angelfishes are particularly antagonistic toward each other, even between species, and it may be hard to

persuade several angels to live peacefully in the same tank, even a large one. Size differences help; a large angel will often ignore a small one of even the same species, but you can't depend on it. The habit of freshwater aquarists of buying a "pair" of fish is to be avoided with most marines. A pair is about the worst you can buy in many cases; one or half-a-dozen of fish likely to tolerate one another is better.

There are other fishes that are far too predatory to be placed in the same tank with any but equally large, tough specimens. Lionfishes, for example, will eat anything they can swallow, even each other if kept hungry. When thinking of acquiring a fish you are not familiar with, make sure of its possible role as a predator. There are tricks by which the chances of such a fish eating smaller tankmates can be reduced, but that is all, not eliminated. If the predator is isolated from future tankmates by a transparent divider and so can see them but not eat them for a week or so, he may leave them alone when the partition is removed—as long as you keep him fed! It is not uncommon for a fish so trained to ignore his tankmates but snap up any new fish of the same species as some of those already in the tank. The behavior of the new fish is probably his clue, possibly sheer visual recognition of the stranger, or both.

Yet other fishes are far too shy and slow to feed to be mixed indiscriminately with a typical marine community. Seahorses and pipefishes that typically eat only livefood, and then quite daintily and at their leisure, do very poorly even when given livefood in a tank with others that snap it up before they get much or any of it. Most mandarins are similar, although the odd one

This butterflyfish is infected with small white spots, *Amyloodinium*, on its body and fin rot infected with fungus on the rear part of its dorsal fin. You certainly don't want to buy such a fish. Many butterflyfishes do very poorly in an aquarium under the best of circumstances.

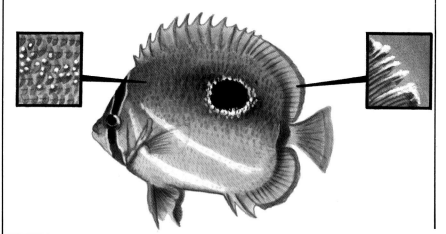

69

gets the message and learns to be sufficiently bold and quick. Even the very occasional seahorse will take frozen brine shrimp or similar dead foods, but this is a rarity. Mostly, instinctive behavior is not broken and the unfortunate mismatched fish just starves to death. Yet placed in a suitable aquarium with fishes of similar habits, these fishes thrive and even breed.

With careful selection initially, it is best to introduce a number of fishes all together to a new aquarium, something that is quite possible if the biological filter has been matured with ammonium chloride or sulphate. They are all new to their surroundings and in no position to guard non-existent territories; also, if a fish can be bewildered, they probably are. They thus have a much better chance of settling in fairly

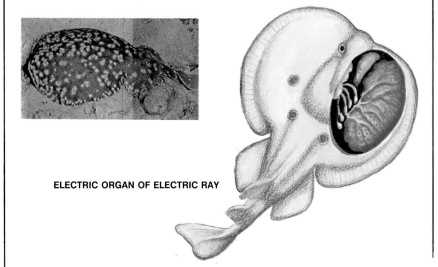

ELECTRIC ORGAN OF ELECTRIC RAY

peacefully, with no worse than a bit of "peck order" bickering. All this is true as long as the tank is large enough and furnished with places to hide and avoid each other if they wish. If the circumstances are such that you cannot do this and must introduce the fishes one or two at a time, purchase the smaller and

more timid ones first, so that they are at home by the time you put in progressively more belligerent fishes in—the old timers will be more able to take care of themselves and the newcomers will be unfamiliar with their surroundings and less likely to get uppity.

Later on, when the aquarium is near full strength and has a balanced community of fishes, the introduction of any but a tough new fish can be disastrous. New small anemonefishes or damsels are likely to be hounded to death by both their own species and most others. The best strategy to minimize the danger is to rearrange the tank so that everybody is put at a disadvantage and there is hope for the new fish or fishes. This may not be desirable (it may upset at least some of the older

inhabitants too much), but it is the best you can do apart from putting up a partition to give the newcomers a space to themselves for a period. I find it difficult to fit a divider into an established, well-furnished tank, and it also interferes with water circulation, so I prefer the rearrangement trick. Even so, I have a large

aquarium at home with mostly long-established angels, damsels, anemonefishes, and wrasses into which I just cannot with safety put any small fishes, except cleaner wrasses, which are of course ignored by instinct. I shall finish up with one large grouper, I expect!

Transfer and Quarantine

The usual method of introducing a new fish to the aquarium is to float in the tank the plastic bag in which the fish will usually have been brought home. Then water from the tank is gradually poured into the bag, removing some of the contents if necessary so as to keep the bag afloat. About a 25% change at each of 3 or 4 steps at 15- minute intervals will do. Depending on the differences in temperature, pH, and salinity, the process takes anything from ½ to 2 hours. It is unusual for a longer period to be necessary, but if there is a big difference in any of the above it would be safer to stage the fish through an intermediate period in a small quarantine tank made up specially for the purpose. Relatively sudden upward changes in temperature can be as much as 5°F, downward ones not more than 3°F. A sudden change in pH is best kept to within 0.2 or 0.3—this is a 60% to 100% change in hydrogen ion concentration, as the pH scale is logarithmic. Similar changes in

Some fishes, such as electric rays (left), are protected from predators. Even a small fish with what seems like a small mouth can swallow a fish larger than itself. This *Synodus variegatus* has the ability to disengage the hinges on its lower jaw to allow the prey to reach its pharyngeal teeth in its throat, where the food fish is chomped into pieces small enough to be swallowed.

There are aggressive fishes that will attack almost any competitor regardless of its size. There are also fishes like the *Pristigenys* shown above, which will only attack fishes it can swallow whole. Note the large mouth!

should be no trouble.

Whether you put a new fish straight into your aquarium or give it some form of quarantine depends on how sure you are of your dealer's practices. If you normally buy from him and have reason to trust him to quarantine his fishes and treat for any diseases that he detects, it will be as safe to float the fish in your exhibition tank right away as to repeat the process all over again. If you have any doubts, it is safer to quarantine the fish yourself. There are several ways of doing this. You may use a small, bare tank with just an airstone and heater, holding from 5 to 15 gallons according to the size and number of fishes. This allows good observation of the fish but doesn't necessarily make him very happy, so perhaps add a bit of coral for him to hide in if he wants to for part of the time, but no gravel or filter. In such a tank you can add medication as a routine or use it only if a disease appears. Some aquarists routinely use 0.15 to 0.30 ppm metallic copper in solution as a cure for the very prevalent velvet and white spot diseases, or some other suitable medication; others wait and see. I prefer the former if I quarantine at all. Some give the fish a short bath of up to one minute in fresh water, taking careful watch to see that it is not unduly stressed, as this procedure kills off many parasites by osmotic shock that takes longer to affect the fish. I am not fond of doing this, as most fishes have been stressed enough without such a shock to their equanimity, and the copper cure is safer. Neither process affects bacterial infections, for which you must watch. It will be necessary to change about 25% of the water in the tank every few days and to discard the water and wash out the tank completely at

salinity should not exceed 0.0025, about 10% change, or the fish may suffer from osmotic shock. A good dealer will usually keep

his fishes around 75°-80°F, at pH 7.8-8.2, and salinity between 1.020 and 1.023. If your tank is within the same ranges, there

the end of quarantine, which should be for 10 days at least. This method is therefore demanding of new sea water or mix and potentially expensive.

Other aquarists keep a small, fully furnished aquarium with an undergravel filter, but not a carbon filter, as this would remove any medication used. Treatment can be as above, since most medications other than antibiotics do not severely affect the biological filter. If medication is only to be applied on detecting disease, a carbon filter can be added and turned off if necessary. This is nicer for the fish or fishes being treated and they can be kept in it indefinitely with only the usual water changes. The new purchases can get used to your usual feeding and lighting procedures and will be in better shape for transfer to the final tank.

Despite all these precautions, no aquarium is going to be sterile, and there will always be the hazard of disease. The final defense against it is the health of the fishes themselves—they have immune systems and will not readily fall ill unless conditions are poor. Immunity in fishes, as in ourselves, may be kept up by small assaults from time to time that keep up antibody production and white cell activity in general. A typical aquarium has in fact several diseases constantly

This large grouper, *Cephalopholis miniatus*, stops by a cleaning station to have the parasites removed from its body.

A cleaner wrasse, *Labroides dimidatus*, cleaning a *Centropyge loriculus*.

infecting the fishes, but mildly, and as long as they are not suddenly assailed by a massive attack of bacteria or parasites they show few symptoms and do not become clinically sick. If experimentally they have been freed of a particular disease such as white spot and then it is introduced, it has been found that the fishes, no longer immune, catch the infection badly and must be treated. In a normal aquarium, a fish that becomes weakened for some reason, such as going on a hunger strike, will often show an infection that the others throw off and thus reveal its presence in the tank. Such a fish is best removed and treated before it *does* assault the others with a massive attack they do not resist successfully. A sudden chill will bring on an attack in similar fashion, this time with many fishes affected because they have all been weakened. Then the whole tank must be treated.

Aggression

Reef fishes are very aggressive and territorial, even when there is no question of predation. Many coral fishes have a quite limited territory that is rarely abandoned and is defended fiercely, in particular against others of the same species. When they mate this behavior is intense, and then all other fishes will be driven off. This sense of home territory is seen in many animals, and typically the owner of it is dominant over others of its kind or even of a different kind that could normally beat it up. The intruder knows that he is trespassing and only in special circumstances will he stand and fight. That is, if he can avoid fighting he will, but what if he can't? Suppose a fish's territory is

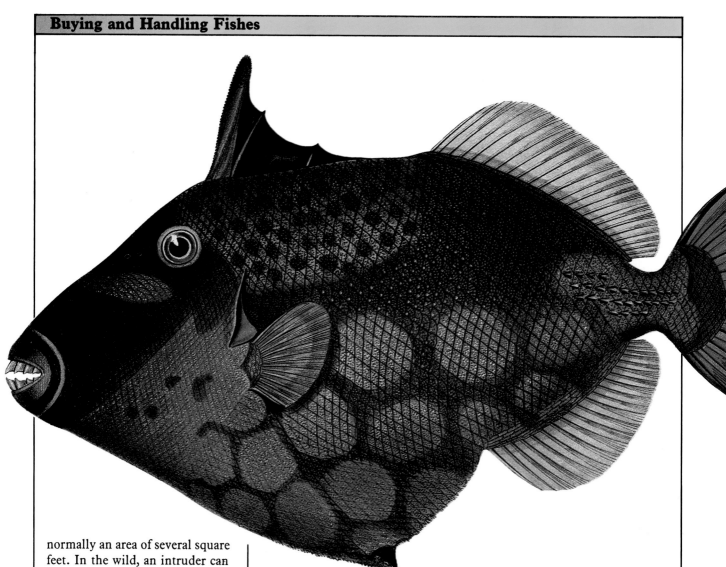

normally an area of several square feet. In the wild, an intruder can move away, but in an aquarium he can't, unless the aquarium is very large. So the battle is on and the intruder, the newly introduced fish, suffers. If he can hide even at the edge of the owner's territory, he may be safe and be gradually accepted; otherwise he'd better be large and tough and not of the same species.

Some fishes tolerate smaller and dissimilar companions within their territories. A large angelfish may not mind sharing with anemonefishes or even dwarf angels, but he is most unlikely to accept another large angel, even of a different species. Yet if they had both been put into a big new tank together, there would have been a good chance of mutual

toleration. Sometimes an upset that is hard to understand occurs in an established community. A fish that has been accepted even for years is suddenly hounded to death by a number of the others—why? It may be perfectly healthy as far as can be seen and yet they suddenly object to its presence. Marine fishes are not like the general run of freshwater fishes; mostly they show more character and seemingly more intelligence, and this makes them fascinating to keep.

It pays to be aware of the degree of aggression to be expected from various types of fishes commonly available.

One of the most beautiful and bizarre of fishes is the Clown Trigger, *Balistoides conspicillum,* shown in this magnificent drawing from Bleeker's *Atlas Ichthyologique.* Note the strong teeth! It attacks everything in the tank.

Outright predators are the lionfishes, groupers, eels, triggers, and anglerfishes. These must be kept with other fishes too large to eat and tough enough to avoid harm from bickering, although the lionfishes and anglers will not attack a fish they cannot swallow. Perhaps the most beautiful and at the same time the most obnoxious of all is the Clown Trigger, *Balistoides conspicillum,* which is safe with nobody and tends to commit suicide even when kept on his own by biting through electric cords or air-lines. Angelfishes most often hate other angels, but there are exceptions, and they are safe with most other species. Dwarf angels harm nobody as a rule. Damsels, including anemonefishes, are very scrappy little creatures and can prove irritating "fin nippers," but they are not as a rule lethal to others. As with angels, the damsels are very likely to hate each other, but many species are safe in large groups, when the aggressive instincts seem to die down. The same is true of anemonefishes. The common *Amphiprion ocellaris* and other small anemonefishes are seen in communities in their host anemones. There is a mechanism, presumably chemical, by which a pair of adult *A. ocellaris* are attended by smaller, younger-looking companions, thought at first to be their young. They are not. Instead, they are of the same batch, remaining undeveloped until one of the "parents" dies. If it is the larger female, the male becomes a female and one of the "young" becomes a male. If it is the male, one of the "young" takes his place. Which one? I wish I knew.

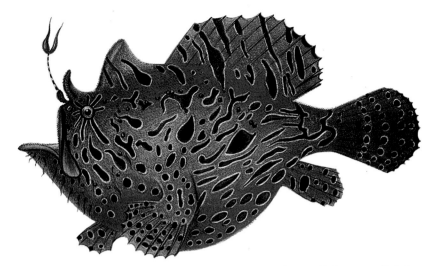

The anglerfishes are characterized by laziness. They wait for a small fish to come by and attempt to eat the "bait." They then slurp in the fish and hardly move for another 24 hours. This is another excellent drawning by Bleeker from his *Atlas Ichthyologique.*

Relatively safe fishes, even when large, are the tangs, wrasses, gobies, blennies, file fishes and butterflyfishes or chaetodons. The chaetodons are also touchy and not for the beginner, often giving trouble over feeding. Wrasses rarely give trouble to other fishes, and the larger ones are death on crustaceans, as are triggers. However, both know the cleaner shrimps and will leave them alone. Cleaner wrasses are good to have in the aquarium, better than cleaner shrimps, as they will not only be left alone but will actively seek out parasites on the other fishes, sometimes nipping a little too enthusiastically and getting chased across the tank. Even so, don't release a cleaner wrasse suddenly into a tank with a large grouper or other predator—he may get snapped up before the big fish recognizes him.

Wrasses do fairly well in the aquarium though the larger ones chomp on coral and graze on crustaceans. The sexes may be differently colored, and males and females have been known to change sex roles. This drawing is from Bleeker's *Atlas Ichthyologique.* The reprint of this series of 10 books is available from the Smithsonian Institution Press, Washington, D.C.

Marine fishes are hearty eaters when given the right kind of food, but they can be very difficult when not supplied with it. Some, such as anemonefishes and damsels in general, will usually eat almost anything; others, such as chaetodons, can be very choosy but can be educated, given time. A few fishes still on the market should never be offered in our present state of knowledge, as they almost always starve to death. Examples of the latter are the Royal Empress Angelfish (*Pygoplites diacanthus*), Meyer's Butterflyfish (*Chaetodon meyeri*), and *Chaetodon ornatissimus*, which doesn't seem to have a common name. Slow feeders can also fare very poorly in a community of other greedy fishes and should be kept on their own or with others of similar habits. These include seahorses, pipefishes, and mandarins.

General Considerations

It comes as something of a surprise to learn that fishes in general are very good at dealing with the food they eat, being in farming language thrifty, meaning that they turn much of it to good use. We convert about 10% of the food we eat into our own flesh when we are still growing, and so do most of the wild mammals that have been studied. Domestic animals bred for thriftiness, such as cattle and pigs, do much better, converting up to 30%, and are beaten by chickens which can manage 35%. The brown trout converts 50% even when on a suboptimal diet, and can probably do better with a richer diet. Salmon perform in a similar manner and have been studied for the effects of temperature. Since they are converting protein to their own flesh, the requirements of this type of food are high. At 45°F

they need 40% protein in the diet, but at 58°F they need 55% to achieve the same weight increase. Food fishes have naturally been most studied, but we may assume that others have similar requirements. Marine fishes need more protein than freshwater fishes, and younger fishes more than older ones, but all need plenty. The percentage of protein discussed is that in the *dry weight* of whichever food is being given. A piece of flesh may have 70% protein after all the water it contains has been driven

off by heating or freeze-drying, but being typically composed of 70-80% water, its wet weight protein content will be only around 15%.

Normal foods will contain carbohydrates and fats, and these are also needed by fishes. Carbohydrates (sugars, starches, etc.) are broken down to simple sugars as in ourselves and provide energy, while the fats utilized by fishes are the polyunsaturated ones; some fishes cannot digest saturated fats, and all have

Pygoplites diacanthus, the Royal Empress Angelfish, should never be purchased as they always starve to death.

One of the most graceful of marine fishes is Meyer's Butterflyfish, *Chaetodon meyeri*. Nobody knows what to feed this lovely fish to keep it alive. The designs on the sides of the body look like something a modern artist might have done; it is, in fact, one of Nature's best camouflage designs.

trouble doing so. Since saturated fats are predominant in red meat of all kinds, too much should not be fed. Instead, the polyunsaturated fats found in fishes and other sea creatures should form the main source of fat in the diet. Mineral and vitamin requirements of fishes have been poorly established. The water-soluble vitamins seem to perform much the same functions as in mammals and birds, and we know that vitamin B1 (thiamine) is needed to assist in the utilization of carbohydrates. Vitamin B6

(pyridoxine) does the same for proteins. Vitamin E (tocopherol) is needed to help in fat utilization, and that's about it. Vitamin C (ascorbic acid) is needed by some fishes, but not by all, as they can manufacture their own. This vitamin is essential for growth, healing, and response to stress. Although some fish livers are loaded with fat-soluble vitamins, it is not certain which, if any, are needed by the fishes, and they may be stored as unwanted byproducts. The trout has been shown to need vitamin D3 (cholecalciferol) for the uptake of calcium, and other studies have shown that an excess of either vitamin D or vitamin A (retinol) is harmful.

Varied Diets

Different species of fishes may feed in nature on quite different substances. There are the predominantly algae feeders, such as surgeonfishes and some angels; predominantly carnivorous feeders, such as triggerfishes, scorpionfishes, and wrasses; and mixed feeders, comprising the majority, that feed on anything from algae and general detritus to various invertebrates including corals and sponges. Some adult angels feed mainly on sponges. Some cardinalfishes and all seahorses and pipefishes feed mainly on plankton or small crustaceans picked from algae. There are many instances of such specialized feeders, some of which stick to their natural diets in captivity, while others are more accommodating and will eat quite differently in the aquarium. It pays, therefore, to acquaint yourself with the natural diet of your fishes and at least start them off with as near to it as you can get, unless your dealer has already accustomed them to something different. Naturally, you won't be able to give an adult angel a continual diet of sponges, but if it hasn't been weaned onto something else before you get it—it should have been in this particular case—do your best, at least at the start. Brown bread looks like a piece of sponge, and some angels love it!

Many fishes change dietary habits as they grow up. Juvenile angels and chaetodons are plankton and algae eaters, also picking parasites from other fishes; later they change over to preferring sponges and corals, respectively. In the aquarium, angels make the change to ordinary fish foods quite readily in most cases, although some are difficult, whereas chaetodons are apt to be more often difficult and some species caught when other than juvenile will not eat

prepared foods or even livefoods other than corals in sufficient amounts to thrive. Fishes that are easy to feed, such as demoiselles, gobies, and blennies, impose only the restrictions that they can get the food down and, as they grow bigger, that it is large enough to interest them. Immatures will gobble up newly hatched brine shrimp, but adults may ignore them, or if they don't, will need vast quantities to satisfy them. Newly hatched marine fishes of practically all kinds need very small first foods such as rotifers or other minute zooplankton, but that's another story.

Variation in diet is needed for quite another purpose, in that within the range of foods that a fish will accept it must be offered as varied a selection as possible. This is because we are not sure that any one food contains all the necessary substances, from proteins, etc., to vitamins and minerals. In addition, if a fish gets accustomed to a particular food, it may refuse anything else, with the consequence that it may starve if the supply runs out. Keep your fishes used to as varied a diet as they will accept.

In a community tank this is easy, as what one doesn't eat another will, but in a specialized tank care will have to be taken to remove uneaten food. Do not be misled by the fact that cats and dogs can do well on a constant diet of a favorite canned or even dry preparation—their diet has been carefully formulated to provide all known necessary nutrients, but nobody can claim to be able to do this yet for a mixed batch of fishes.

We shall now look at the various foods you can offer your fishes, with some recommendations about feeding particular species.

Natural Plant Foods

A good aquarium will have its own growth of algae, both the encrusting types that cover the rocks and coral and the sides and back of the tank and the filamentous types or others that look like higher plants. Fishes that are algae eaters will be seen cropping both types, some in a virtual continuous feast if there are enough algae for them to do this. This algal growth is purifying the water and feeding the fishes at the same time, and it should be encouraged as a sign of well-being in the tank. Some aquarists clean it off and bleach the coral far too enthusiastically and deprive themselves and their fishes of one of nature's best helpers, as well as a decorative addition to a mature-looking aquarium. Indeed, it is a good thing if you have to cull some of the algae yourself as you are then removing some of the products of fish metabolism from the tank, which of course doesn't happen when it is eaten by the fishes. If you keep surgeons and some of the angels, you will be wringing your hands as they eat up any higher algae you may be trying to cultivate, but with many species

This beautiful butterflyfish, *Chaetodon ornatissimus*, certainly one of the most beautiful fishes in the ocean, has proved impossible to keep in the aquarium because it slowly starves to death.

you can grow algae such as *Caulerpa* species and add a very attractive item to your aquarium. All that is needed is adequate light and perhaps one of the commercially available algae feeds—a mix of iron and other

are going to be eaten soon anyway.

Many of the same livefoods fed to freshwater fishes are quite suitable for marines. They will live long enough in salt water to be eaten; some will live for hours or days if given the chance, such as mosquito larvae and bloodworms. These foods are available in most pet shops or can be collected by you. Others can be cultured—microworms, Grindal worms, whiteworms, and earthworms, for example. Do not feed such worms frequently, as they are rich in saturated fats and form an unsuitable regular diet for marine fishes; regard them as treats.

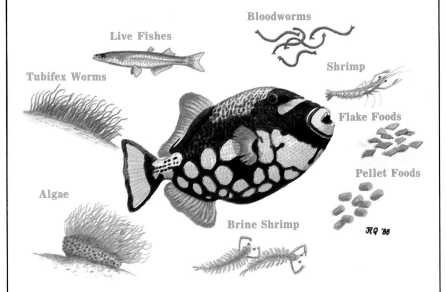

Bloodworms
Live Fishes
Shrimp
Tubifex Worms
Flake Foods
Pellet Foods
Algae
Brine Shrimp

trace elements that encourage plant growth.

Depending on the balance within a tank, natural algal growth may or may not be sufficient for the fishes. You can substitute with various natural materials from dried seaweed, available in health stores or Oriental food shops, to chopped land plants. The best of these are lettuce, spinach, and any green, soft-leaved plants eaten by ourselves. The best preparation of these for the aquarium is to deep-freeze the leaves after chopping them to a suitable size. If you prefer, they can be stuffed in a jar and chopped up afterward. Fishes can learn to eat fresh leafy material, but they normally take it more avidly after it has been frozen. Perhaps it then looks more like algae; it is certainly softer and more glutinous, but whatever the reason, they like it! As the cells have been disrupted it will be more digestible as well.

Essentially similar plant material, whether originally algae or not, will be present in some prepared foods.

Live Animal Foods

With the abundance of frozen, freeze-dried, and canned foods available, it is not strictly necessary to feed livefoods to your fishes except when raising fry, and you won't be doing that very often! Nevertheless, a feed of live creatures now and again is much appreciated by most fishes and probably helps to keep them in better health. With the exception of brine shrimp, it is best to offer livefoods from freshwater sources, eliminating the problem of introducing parasites into the tank. If you do make a collection from the shore or the ocean of goodies for your fishes, sterilize it as far as possible by dunking it in fresh water for a minute or two, as it doesn't matter if harm is done to the creatures it contains—they

Insect Larvae

Mosquito and other insect larvae are valuable live foods as many survive in salt water until eaten and can therefore be fed fairly freely. As they breathe air, they do not compete with the fishes for oxygen. When collecting them it is usual to avoid including the larvae of predators such as dragonflies or water beetles, but even that doesn't matter when they are going into a marine aquarium. Mosquitos may lay their egg rafts in brackish or salt pools, but it is best to avoid such sources or to store the larvae in fresh water

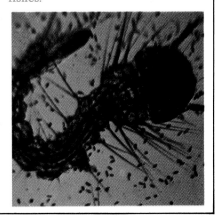

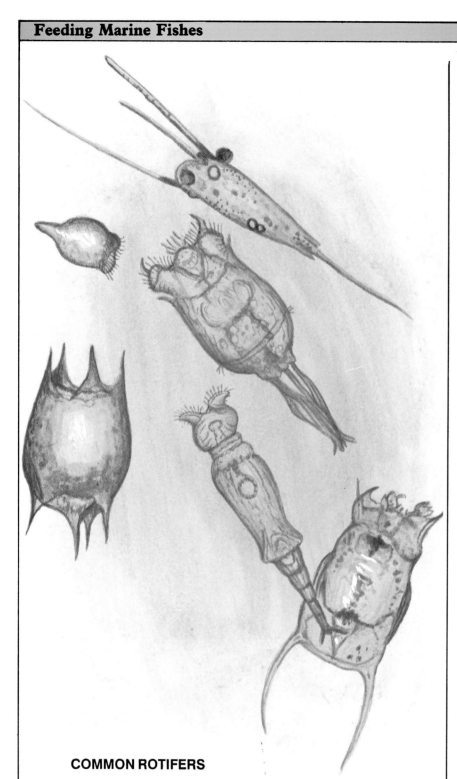

COMMON ROTIFERS

The usual live foods offered to freshwater tropical aquarium fishes may safely be given to marine fishes. The idea is to feed them rather sparingly as most of the freshwater rotifers do not live too long in marine waters. There are many types of rotifers to be found in freshwater pools worldwide. The drawing shows some of the usual forms. Good petshops usually have on offer a constant supply of live foods for aquarium fishes. It might be a good idea to try them all if they are small enough for your fishes.

before use. The small, dark brown egg rafts hatch into hundreds of tiny "wrigglers" that hang under the surface of the water and grow, depending on species, to up to nearly 1/2" long during the next eight or nine days. These turn into pupae, harder and rounder than the larvae, that are eaten by most fishes but soon turn into unwanted adults, so beware. The larvae and pupae are collected from still pools with a fine net and can then be washed and sorted for size, if necessary, with domestic sieves. Store them in closed jars of water in the refrigerator, *not* the freezer, and they will keep for several days without losing much nutritive value.

The bloodworm (actually the red larvae of *Chironomus* and many other genera of the family Chironomidae) is found in the same localities as mosquito larvae but is less often encountered. The larva of a gnat, it is larger than the mosquito larva and is a fine food for medium to large fishes. Bloodworms are found deeper in the water than mosquito larvae and are more difficult to collect clean. In contrast to the deep red bloodworm, the glassworm, the larva of *Chaoborus*, another gnat, is colorless and is found in cold weather. It is an equally good food.

The mealworm, the larva of a flour beetle, is cultivated as bait and is another fine fish food. Feed it whole to large fishes and chop it up for smaller ones. Blowfly maggots are similarly appreciated and can be cultivated if you choose by placing some flies in an enclosed vessel with a piece of meat. They will lay their eggs and the maggots will grow on the meat—all very smelly and best done outside. I repeat the warning that these live foods are for occasional use only—not more than once a week.

Worms Earthworms, fed whole or chopped up, are another good occasional food. They can be purchased, cultured by you, or coaxed from the lawn by pouring a solution of 1 grain per gallon of potassium permanganate over it, when they will emerge if there are any there. Store them in leaf mold, slightly damp. If you are sensitive, they can be killed in boiling water and then chopped, but fishes seem to like them better fresh. Fresh from the earth, the worm is liable to be full of dirt, so strip the contents of its digestive tract by pulling it through the fingers.

Tubificid worms of many species, all sold under the name of "tubifex", are red to brown in color and up to 3″ or 4″ long. They are from filthy sources—slow-moving, polluted water with a muddy bottom in which they construct tubes, waving their tails in the water to obtain such oxygen as they can get. You can collect them or buy them. Unless you enjoy paddling in sewage, it is best to do the latter. The worms must be cleaned by keeping them under a dripping tap until there is no smell, no cloudy water, and muck entangled with them. Break up any solid bunches and wash well, as they often die in the middle of

such clumps. They may then be stored in the refrigerator, if your spouse or parents aren't looking, and used with a further rinse over the next few days. Tubificid worms are very fatty and must not be fed in excess. In fact, I only use them to coax finicky eaters such as chaetodons, which will often take them stuffed into dead coral skeletons when they won't look at any other food. Perhaps they look like coral polyps to the fishes.

Whiteworms, *Enchytraeus albidus*, and Grindal worms, a smaller species, are good occasional foods. The former grow to about an inch and the latter to half that size. They are both roundworms related to tubificids but are found in damp soil or anywhere that is dark, damp, and has decaying matter of some description. They may be cultured in soil to which milk and any porridge-like food have been added. The box should be covered by a glass cover (to keep in moisture) and then with an opaque one to keep them dark. Place the enriching food in pockets on top of the soil. Start

with a culture from a dealer or friend. Worms will collect around the food, under the glass, so that you can see how things are going. Wash the worms in a very fine net before feeding to the fishes. White worms do best in a cool place, not over 70°F, while Grindal worms like it warmer, at 70°-75°F.

Microworms, *Anguillula* species, are very small, up to a maximum of 1/10″, and are of questionable use for marine fishes. It seems possible that they could be used for feeding marine fry, as the microworm young are minute, but I have not heard of this being done—perhaps they would die too quickly or not be taken anyway. They are a valuable first food for freshwater fry.

Brine Shrimps These tiny crustaceans (*Artemia*) have been used as a livefood for decades in the newly hatched form and are frequently available as adults today. Frozen, they are the most common fish food other than flakes. They made a tremendous difference to the raising of

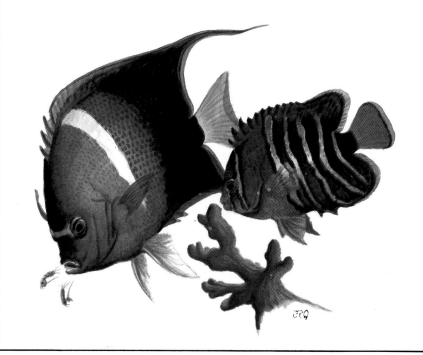

USING BRINE SHRIMP. 1. Brine shrimp eggs, nauplii, and adults. 2. Petshops sell the necessary salt, pump, and eggs from which to hatch your own live food when you need it. 3. Add salt, water, and eggs. 4. Wait a day or more for the eggs to hatch. 5. Take out the airstone and net the brine shrimp. 6. Wash them under fresh water. 7. Feed them directly to the minireef aquarium. 8. You can grow them to adult size in a large tank, in the sunlight, with a little brewer's yeast. Your petshop has all the details. Drawn by John R. Quinn.

freshwater fry, and it is a great pity, that of aquarium fishes, only newly hatched seahorses and some pipefishes can be raised on them; most other marine fry need smaller organisms at first. The dried eggs accumulate naturally in heaps or layers around the edges of evaporating pans in salt works and around some salt lakes. They are very small, a pinch containing many hundreds or even thousands of eggs, and are packed under vacuum for long term storage, when they will last for many years. Small vials or packets are available from retailers and can be hatched in sea water or somewhat more dilute salt solutions. The adults live in a much denser brine, but the eggs presumably get washed back into diluted waters when it rains and have thus become adapted to hatching in such conditions. Brine shrimp eggs are also available in shelled form, although naturally more expensive, and can be hatched in the marine aquarium directly or

A beautiful reef scene in which can be seen sea fans, sea whips, corals, sponges, and other invertebrates as well as fishes. Most prominent of the fishes is the very popular *Heniochus acuminatus,* commonly called the "poor man's Moorish Idol." Photo by Cathy Church.

in sea water that can then be poured straight into the tank. Otherwise, more involved hatching procedures must be followed.

If only a small quantity of newly hatched brine shrimp is needed, we use the following method. Sea water, a mix, or two heaped tablespoons of common salt per quart of tap water are placed in a shallow pan such as a photographer's developing dish. When it is still, not more than one-half teaspoon of eggs per gallon is sprinkled carefully over the surface and preferably covered—light or dark doesn't matter. For small quantities, the little spoons sold in sets usually rated ¼ teaspoon, ½ teaspoon, etc., are very useful. A temperature of not less than 70°F, preferably nearer 80°F, is needed for hatching. At 80°F the eggs will hatch in 24 hours, in 48 hours or more at 70°F, and not at all if much colder. Spread the eggs as evenly as possible and be careful not to sink any of them. As they hatch, the nauplii, young shrimp larvae, sink down and remain on the bottom for about eight hours. Later, they swim up into the water and become phototropic—attracted to light. As they will not all hatch at exactly the same time, it is best to wait for this stage, say two days from the start for safety, three days if not above 70°F, and then to collect them.

To harvest the nauplii, dip a flexible siphon tube below the surface of the water and run it either into a collecting vessel or onto a very fine filter cloth, according to whether you wish to pour the nauplii straight into the aquarium or to discard the hatching fluid—the latter is best, as even if it is sea water it will be a little contaminated. With care, practically all of the nauplii can be collected free of egg-shells, which will still float on the water and be left behind. The hatching fluid can be re-used if necessary once or twice.

When large quantities of shrimp are required, less bother and less water are involved if the next method is used. Take one or more gallon or larger bottles and fill to about ¾ with the hatching medium made up as before. Put in an airstone and up to a teaspoon of eggs per gallon. Turn up the aeration so that the eggs are whisked around briskly, then after the hatching period turn it off and wait for 15 minutes. Some egg shells will float and some will sink to the bottom, but the nauplii will be free-swimming and can be siphoned off as before. If they seem too dispersed and difficult to harvest free of shells, take advantage of their phototropism and direct a beam of light across the middle of the bottle, where after a short period the great majority are easily caught.

You can grow your own brine shrimp quite easily if you think it worth the trouble, as illustrated by the little kits sold as a novelty in which "water monkeys" are cultured. The nauplii will live on their stored yolk for a few days but must then be fed. They feed in nature on algae and bacteria, for which baker's yeast can be substituted. But first, transfer just a few, about 500 per gallon if

Two types of the popular *Caulerpa* algae

Because of the intense lighting of most mini-reefs, algae grow well in them, and you may even have a surplus to feed the fishes. Although encrusting algae of various types are most common, it is no longer considered difficult to grow true seaweeds such as *Caulerpa* and similar forms. Starter cultures of several types are now available commercially and often can be bought at your local aquarium shop. The essential trace elements necessary for successful algal growth are also available and now allow almost any tank to contain a good growth of beautiful and useful seaweeds. Of course, if you keep fishes with the algae you can expect some problems, as many types of fishes will eat the algae as they appear.

you wish to raise them to adult size, more if not, to a denser brine. This is, for San Francisco shrimp, made up of 10 oz of common salt (NaCl), 2 oz of Epsom salts (MgSO$_4$), and 1 oz of baking soda or sodium bicarbonate (NaHCO$_3$) per gallon of tap water. This brine is about twice the strength of sea water and much more alkaline. Stir a few pinches of yeast into the brine and keep it suspended with brisk aeration. Feed more yeast as the brine clears and keep covered. At 70°-80°F the shrimp will become adult in six to eight weeks, when they are about ⅜" in size. If left alone, they will breed and keep up a continuous culture if not all are removed, as they produce eggs that will hatch without drying first.

Other types of brine shrimp, such as those from Utah, need a different brine, containing borax, for example. Information about their requirements is often given on the container or can be obtained from the suppliers. A brine concentrated from sea water will often suit them, or try making up a double-strength mix of your usual salts. Some types of brine shrimp will even grow up and breed in straight sea water.

Other Crustaceans Brine shrimp are soft-bodied, having little in the way of the hard carapace characteristic of other crustaceans. For this reason they are less stimulating to the gut of the fish that eats them and should not be fed exclusively. Small crustaceans collected from either fresh or salt water are preferable in many ways, although those from the sea should have a wash in fresh water prior to being fed to the fishes. Freshwater crustaceans do not carry parasites of marine fishes and so are safer.

The water flea *Daphnia pulex* swarms in outdoor pools in the

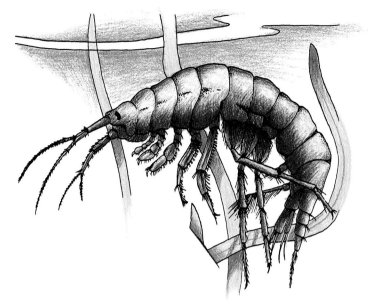

Amphipods are small crustaceans commonly found in the bottom of both freshwater and marine habitats. They are eaten by many fishes, both large and small, and are easily raised in a separate tank by just introducing a few egg-carrying females into a layer of detritus. The marine forms are usually harmless, but it might be safer to use freshwater species in the mini-reef to prevent any problems.

If you are able to collect in an area where the water is pollution-free, you should be able to obtain quantities of various planktonic plants and animals by using a long, conical, very fine mesh net called a plankton net. These minute creatures are excellent food for many young and very small fishes, and often they are almost the only food that will be taken if you are trying to raise larvae.

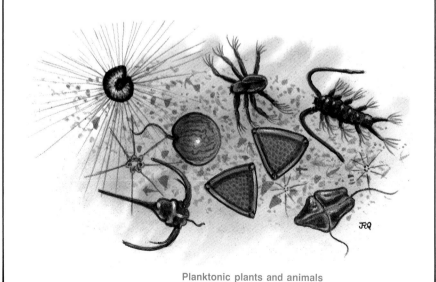

Planktonic plants and animals

summer, although it does not like very hot weather. They can be collected in quantities with a fine net, occurring as red, green, or yellow clouds in the water, depending on their principal sources of food and the particular strain of flea. Wash them well and keep them in cool storage. In the marine aquarium they live amply long enough to be eaten if not too many are offered. They sometimes occur in such numbers that if placed in a usual collecting vessel they might suffocate; this can be avoided by laying them onto flat trays of soaking wet cloth—they will survive for hours as long as you keep them wet. They can be bred, but only in quite large aerated tanks or tubs kept fairly cool and holding 50 gallons or more. Feed them on liver powder, dried blood, and other material rich in protein.

Other small freshwater crustaceans are *Moina*, *Diaptomus*, and *Cyclops*, the latter being a pest in freshwater aquaria but of no danger in marine ones. *Gammarus* and *Asellus* are larger, shrimp-like creatures found in streams as well as ponds, and also breedable, but in cold water, otherwise as for *Daphnia*. *Hyalella* is a warm-water breeder, common in the U.S.A. but not in Europe.

A great variety of small crustaceans can be collected with a fine net in tide pools, eel-grass beds, sandflats, and even the ocean. These will be a mixture of amphipods (small laterally compressed relatives of the sand hopper), copepods (crustaceans that, like *Daphnia*, swim by jerking their antennae), phyllopods (fairy shrimp, relatives of the brine shrimp), isopods (relatives of the woodlouse or pillbugs), and above all decapods (shrimps, crabs, lobsters, etc.) in their small sizes or stages. There are so many thousands of species in total of these crustaceans that it is pointless to try to describe them in detail. How far up the scale in size you care to go when collecting them depends on the fishes you have. Quite large shrimps and crabs will be much appreciated by wrasses, triggerfishes, and lionfishes, but will settle down unharmed in a tank of smaller fishes like anemonefishes and damsels. Some of the copepods and isopods are parasites of fishes, but it is unlikely that you will collect these from the sources mentioned, particularly if you are collecting from temperate waters and your fishes are tropicals.

Daphnids are suitable food for many fishes and invertebrates and provide the roughage necessary for digestion. *Daphnia* and its relatives, both marine and freshwater, are sold in many aquarium shops both as live food and in freeze-dried form.

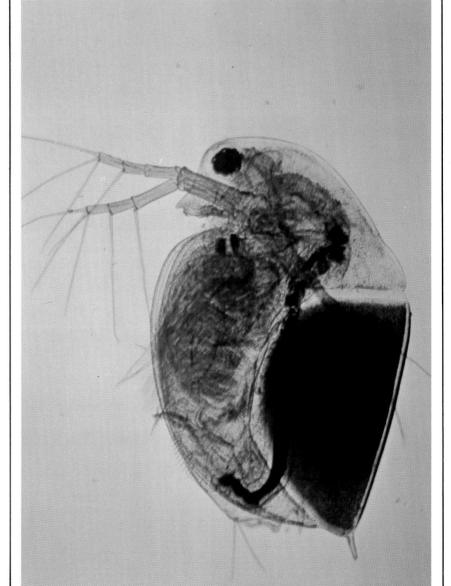

Plankton from the ocean is mainly the larval stages of crustaceans, plus lots of other small critters and some algae. Krill, which look like small shrimp, are also sold as plankton, so make sure which you are getting if you buy frozen plankton.

Mollusks Live mollusks are perhaps a bit risky as foods, particularly the filter-feeders, but they are much appreciated by many fishes. In fact, I found that the only food accepted in any quantity by my favorite and very finicky Royal Empress Angelfish (*Pygoplites diacanthus*) was freshly opened live mussels. She lived and grew for a year on it, only to

be bullied to death by a large Koran Angel before I realized what was happening. She went off her food and nothing saved her, even transfer to a tank on her own. Live scallops, small clams, and oysters can be cut open in similar fashion and in my own experience have not caused trouble, despite the fact that they are filtering off potential disease-causing organisms.

Fishes The fry of freshwater fishes of all kinds and the young of livebearers are much appreciated by the great majority of marine fishes. If you are a freshwater aquarist as well as keeping marines, give all your culls and unwanted spawnings to

the marine fishes as they form a rich diet of goodies such as vitamins and essential amino acids sometimes lacking in other livefoods. The predatory fishes, lionfishes, anglers, groupers, etc., thrive on as frequent a feeding of live fishes as you can supply, even when they have been trained to accept other foods. Substitute foods rarely contain the proper mix of dietary components supplied by whole fishes. Be careful not to feed any of the predators with larger fishes than they can comfortably swallow, or two unfortunate consequences may follow. First, the prey may be regurgitated half-digested and foul the tank; second, particularly with lionfishes, the predator sits on the bottom looking sorry for himself and breathing rapidly with frequent gulping motions. This may pass, but in some cases the fish seems unable to reject his excessive meal and actually dies.

Processed Animal Foods Of the worms mentioned as livefoods, the only ones offered in prepared package form are tubificid worms, which come freeze-dried. We must suppose that they were adequately cleaned before being processed. Certainly no harm seems to come from an occasional feed of them, and they are much appreciated by most fishes.

Crustaceans Deep-frozen or freeze-dried adult brine shrimp are a favorite offering to all small fishes, freshwater and marine. As with all dry foods, freeze-dried brine shrimp is very concentrated. If gobbled up by the fishes it may swell in the

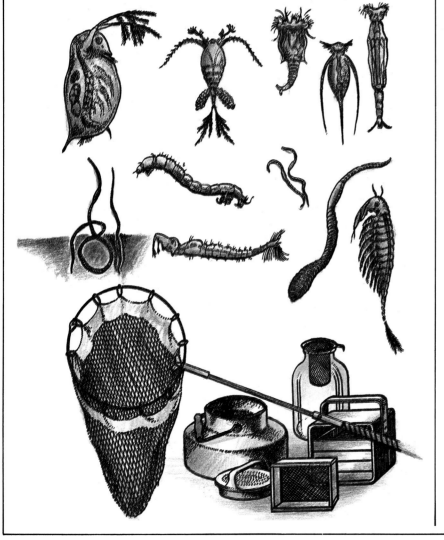

Very little equipment is necessary to collect live foods suitable for almost any fishes and invertebrates you keep in the aquarium. Most are also available at your petshop.

Manufacturers have developed specific foods including frozen foods like those shown here for many of the reef animals that are kept by aquarists. Be sure to check them out at your pet store so that you can offer your charges a proper diet. Photo courtesy of Ocean Nutrition.

When feeding food you have collected yourself, be sure to clean it thoroughly, removing all extraneous animals and possible parasites. One of the dangers of feeding collected animals such as these amphipods is that parasites might be introduced to the tank.

stomach and give trouble, so don't dump too much onto the water at once, or soak it in a little water first—fresh or salt, it doesn't matter. If you find that all of your fishes like it soaked, well and good, but some seem to prefer the dry foods before they have had a chance to absorb much water. The deep-frozen variety is just like the living article as long as it has not been poorly handled, either in the processing or afterward. If it has been poorly prepared or thawed and frozen again one or more times, the body and even the cells of the shrimp are rendered porous and leak much of the fluids plus nutriments they contain. The result can be a package full of skeletons and a brown fluid that does nobody any good. Most of the hard-shelled crustaceans can withstand such treatment, as even if their cells are ruptured it is mostly all contained within the watertight external shell and is still eaten by the fishes.

Other shrimp are a mainstay of many marines in nature and also in captivity. They also come frozen or dried and can be

obtained in a variety of sizes to suit various fishes. Mysid shrimp are no larger than brine shrimp and are hard-shelled, forming a very good diet for smaller fishes. Shrimp of various larger sizes, usually incorrectly offered as "krill" can be fed to larger fishes or, in the case of the freeze-dried ones, crumbled for the others. Shrimp or prawns for human consumption can be fed, preferably raw, but cooked ones are accepted quite readily and are perhaps safer from the point of view of infection. However, the danger from raw materials would seem to be remote. The larger raw crustaceans should be peeled before use, as except for some of the predators that know how to deal with them, most fishes cannot cope with whole unpeeled ones.

To the shrimp can be added portions of crab, lobster, or any other edible crustaceans, raw, cooked, canned or frozen. All are good for marine fishes and only need to be chopped up or grated to sizes suitable for your fishes.

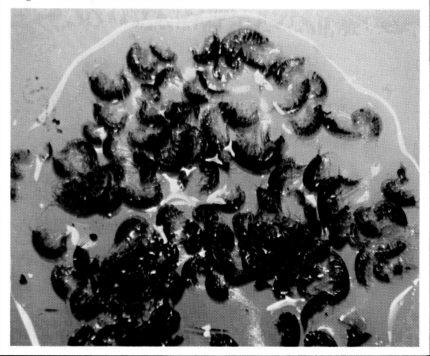

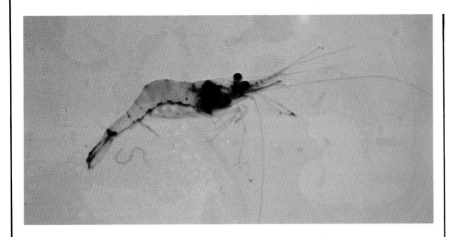

Small shrimp such as the *Palaemonetes* above are readily collected or purchased and are relished by the majority of aquarium fishes. Although mussels (bottom) and other mollusks are readily obtained, they often are contaminated by pollutants and pose some risk in the aquarium.

Mollusks The same mollusks that can be fed live may be offered after freezing or in the half-dead condition seen in fish markets. The filter-feeders are probably more dangerous when moribund and should be washed well in fresh water, when they can be given in the shell or chopped up. In addition, octopus and squid may be chopped up or shredded in a food processor and will keep for weeks in the deep freeze. You can make up a mix of chopped squid, scallop, and krill of various sizes and freeze it in ice cube trays for later feeding. Sometimes add some chunks of shrimp or prawn, the grand mix providing food of different sizes for all comers. It is best to thaw it out before feeding, but it is quite in order to float a cube in the tank and let the fishes pick at it as it thaws—as long as everyone gets a fair go. With most of my own tanks, I find that this is not always the case, but if I thaw it and throw the lot in at once even the timid customers get some.

Fish Fish from the fishmongers, deep-frozen from your dealer or the market, in can, or even dried and reconstituted is excellent in moderation. Whole small fishes fed to predators or sliced up for the others are best, as then they get the whole works, including liver and intestines and bones, which is better for them than just the muscle flakes. In fact, fish muscle has been shown to have vitamin-destroying properties that make it unwise to feed much of it unaccompanied by the rest of the fish. Ocean fishes such as cod and mackerel are said to help prevent strokes and heart attacks in humans, more than do inshore or freshwater fishes, but I don't know if they are of particular benefit to other fishes—perhaps to whales or dolphins?

Red Meat Beef heart, ham, and other meats are often recommended as good fish foods, but be very cautious about their use. A very occasional feeding may not matter if you can do no better, but these are not natural foods to any but sharks and a few large predators, and are not eaten very often even by them. The animal fats they contain are mostly saturated and solid at aquarium temperatures and are indigestible to fishes. Chicken is better, less fatty and with some polyunsaturated fats as well, but again, it is not to be fed very often. The best thing that can be said for all such meats is that they usually will not carry diseases or parasites to the fishes, even when given uncooked.

Dry Foods Dry foods in the form of flakes, granules, or pellets are a great attraction to the aquarist, especially when they promise a complete diet, perhaps

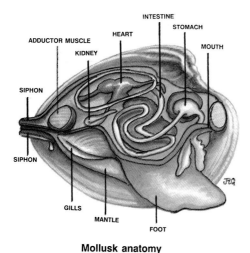

Mollusk anatomy

The clams and mussels are bivalves, mollusks with two shells, a large foot used for burrowing, and no obvious head. The large gills concentrate many pollutants from the water.

go further than a feed of dry food once per day with another of livefood or frozen or canned prepared food each day as well.

Flakes When buying flakes, prefer those made specifically for marines, as those intended for freshwater fishes are not always suitable. Go for a good mix of animal and vegetable material, best with added vitamins, and for not less than 45% protein. Remember the high protein requirements of marine fishes. Most flakes are colored. In most cases this means nothing; it is to attract you, not the fishes. However, there are colored flakes available in which the color does indicate the content and to an extent the purpose of the flake— to feed vegetarians, to feed carnivorous fishes, to stimulate color in the fish, etc. Go also for a chunky, crisp flake that isn't accompanied by a lot of powdery residue and isn't too thin to crumble if necessary without just becoming dust. Although the protein and other contents of the flake are specified, there is some variation in the way they are calculated. The best you can do is to buy a reputable brand and trust the manufacturer not to be too optimistic in his analyses. The same applies to the constituents, that may be specified as shrimp, crabmeat, fish roe, oatmeal, etc., etc. but with never a hint as to how much of each. There may be 10% shrimp, 1% fish roe, and 50% oatmeal—who knows?

Keep the flakes well sealed and in a cool, dry place. Another thing not specified is a "use by" date, necessary for some human foods but apparently not for fishes. Buy from a shop with a good turnover and hope for the best, and don't buy too much at once. It is usual to feed flakes sparingly since they are

with added vitamins. Many a freshwater fish lives on nothing else and doesn't do too badly. In the bad old days, the staple diet of many goldfish was the white wafer, composed of rice flour and egg white, on which some at least apparently survived! Other foods were ground-up dog biscuit and granular efforts made of dried insects, fish roe, flour, meat, and other similar ingredients—

probably better food than the wafer or biscuit, and not very much different from some foods offered today. Marine fishes need better diets than any of these, although it does no harm to let them have an occasional meal of dry food. In fact, it is advisable to accustom them to taking such foods in case you find yourself with no alternative—which is pretty unlikely, however. Never

Although these file shells, a relative of the scallops, are readily eaten by many fishes with heavy teeth, their bright colors make them more useful as display specimens than food.

concentrates, and fishes can gorge themselves with dire consequences if allowed to do so. The flakes then swell in their tummies and can cause great distress. Some aquarists soak the flakes before feeding, which has the advantages of avoiding post-prandial swelling and some will sink down immediately to feed bottom-dwellers. Otherwise, they float and are mostly eaten at the surface. I give a swirl to the water as I feed, which helps to produce the same effect, and some at least of the flakes rapidly absorb water. A good flake doesn't cause cloudy water if left uneaten, so a little left over is in order, to be eaten soon anyway. Don't leave much, though, as no food should remain in the tank for long or it will add unnecessarily to pollutants.

Granular and Other Dry Foods Flakes are made by drying a gelatinous mix over rollers and cannot readily contain ingredients that are not reduced to a very fine state. Granules can offer a chunkier diet with recognizable bits of insects: *Daphnia*, spinach, fish roe, and so forth. They are suited best to the larger fishes but can be ground fine enough to be fed to even tiny ones. Other dry foods

are made from single species, usually freeze-dried, such as bloodworms, *Daphnia*, krill, brine shrimp, and tubificid worms. These are all concentrated foods, and the same constraints in using them apply as to flakes. They are much appreciated by most fishes and in some cases offer the highest protein content of anything available, up to 65 or 70% in the case of krill. Avoid the old-fashioned "ants eggs," that are the pupae of ants, not their eggs, and are of poor nutritional value. Avoid also the temptation to feed the cheaper foods, pellets or flakes, intended for pond fishes. They are very low in protein, usually lower than they should be

Left: Goatfishes, such as the *Pseudupeneus maculatus* shown here, are bottom feeders, using their barbels to detect food lying on or just under the substrate. Needless to say, food for these fishes must sink to the bottom where they can utilize it. Photo by Courtney Platt.

A tremendous variety of dried and processed foods is available at your local pet shop. These foods allow every aquarist to provide a broad range of foods to his animals, many of which might not be available in any other form. Most modern flake and pelleted foods have added vitamins and minerals to assure a balanced diet as close as possible to the natural regimen. The convenience of processed foods is of course far greater than of live foods. Photo courtesy of Hikari Fish Foods.

even for coldwater fishes. I feed my pond fishes marine flakes or bloodworms, since I don't have so many that it would be too expensive, and boy, do they breed and grow fast!

Homemade Mixes Recipes for all sorts of homemade mixes are given in fish books, mostly intended for freshwater fishes. They may be cooked, dried, and then ground up; mixed raw and frozen in ice cube trays; made into gelatin cubes; or even mixed with plaster so fishes like parrotfishes and some wrasses that normally chew at coral or scrape things off rocks can have a natural-seeming diet. It is only worth going to a lot of trouble if you have a lot of fishes, otherwise it is not very expensive to use commercial preparations supplemented with tidbits now and again from the kitchen—bits of shrimp, crab, or lobster, some frozen green lettuce leaves, a little raw shellfish or raw fish. Be careful not to feed oily canned foods; wash them well first and then they are O.K. Perhaps the foods most lacking in common diets, particularly if you keep tangs or angelfishes, are the vegetables. Good flakes have sufficient amounts, but most of the frozen foods are meaty and must be supplemented with a generous amount of vegetation unless your aquarium itself provides it. Freeze it first for increased edibility.

I feed a mixed lot of marine fishes and invertebrates and find that the best way to prepare my own mix is as follows. Squid, octopus (preferably whole), scallops, mussels, small baitfishes, shrimp, crab, any small crustaceans, any convenient vegetables, sometimes a little wholemeal bread, form the basis of the mix. Using whatever is at hand, the big stuff is chopped

into small pieces averaging about ¼″ x ¼″ x ½″, anything much smaller is left as it is. The plant material, usually lettuce, and any bread are chopped quite small. Enough of the mix is made to last about 3 weeks and is frozen in ice cube trays. All shells except those of the smaller, untouched crustaceans are discarded. A human vitamin capsule containing the normal amounts of water and fat soluble vitamins recommended as a dietary supplement (no megadoses!) is added per approximately 1 lb of total mix.

When feeding, an appropriate amount is thawed out, which I find is around one cubic inch per

20-30 medium-sized fishes varying from 1½″ to 6″ in length, once per day, often with a later feeding of flake or dried food. Depending on the actual contents of a particular tank, the mix may be chopped fine or left virtually as it is. Usually at least part of it needs a further mincing, but for large fishes and for feeding anemones, large hermit crabs, and some starfishes the bigger pieces are offered whole.

Feeding a tank of mixed fishes and invertebrates needs strategy if everybody is to get some of the mix. Anemones with anemonefishes are particularly difficult as they cannot be fed at night, when most fishes are resting and will not bother them,

Anemones are often hard to feed because the anemonefishes will steal the food before the anemone can utilize it. It is necessary at times to feed the fishes first and then concentrate on the invertebrates. Photo by M. P. & C. Piednoir of *Amphiprion sandaracinos.*

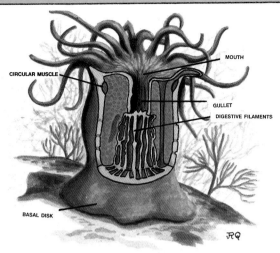

Internal structure of a sea anemone

The simple internal structure of the anemones makes it possible to easily stick a small chunk of suitable food into the mouth of the animal and have it accepted.

CIRCULAR MUSCLE
MOUTH
GULLET
DIGESTIVE FILAMENTS
BASAL DISK

because their fishes are nestling in them and it doesn't seem wise to disturb them. Feeding them by day is difficult since many fishes, including the anemonefishes, will steal the food before it is safely inside. If there are crustaceans present they will often learn to fish down inside the anemone and get the food even when it has been ingested, as they are immune to its sting, although smaller ones can be caught and ingested themselves. So, with anemones and shy feeders like some of the gobies and blennies, I put some food in at one end of the tank and feed the difficult ones at the other end, thrusting chunks down into anemones, which they soon learn to accept. Then, if necessary, the process is reversed. Otherwise, the best way to make as sure as possible that all of the fishes have a chance is to feed the whole meal at once and see that it is swirled around the tank. Don't worry about transient cloudiness; if your aquarium doesn't clear in 10 to 20 minutes there is something wrong. The finer cloudy particles feed many types of invertebrates, but if these are not present the undergravel filter should take care of the situation.

Specialist Feeders Although various seastars, crustaceans, and other invertebrates than anemones or filter-feeders may get some of the daytime feeds, often they are half-starved unless you feed them at night. It is a good practice when such creatures are present in a tank to drop in some small pieces of shrimp, clam, or such well after dark when the fishes are inactive but the invertebrates will be found wandering around searching for any left-overs and usually not finding very much. There is no need to do this every night; just twice a week or so is enough to make sure that they are adequately fed.

Filter feeders, clams, mussels, scallops, sponges, many corals and anemones, and some starfishes ingest fine food particles or living plankton as they pass sea water over their gills or their surfaces or in some cases sift the detritus or sand, if present. Sea cucumbers in particular do the latter. Some get enough to live on from the smaller particles of left-over food, but others require purposeful feeding. A favorite food for these is newly hatched brine shrimp, which can be fed by a small

syringe or kitchen baster directly onto them. Other suitable but preserved foods are available commercially, usually labelled simply "invertebrate food", which are similar to the liquid suspensions sold for freshwater fry. Indeed, fry foods can also be used for invertebrates; it is best to mix them with sea water first or they will float up out of reach. Some writers advise removing corals or tubeworms to a small vessel for feeding; while this does ensure that they get a good feed, it is not always practical.

As for fishes, there are several groups of specialized feeders that are worth a mention. Some chaetodons are very hard to feed but are trainable if you go about it the right way. They will often take material, particularly tubificids, stuffed into the

With a simple trick, several otherwise difficult butterflyfishes can be made to eat tubificid worms—just stick a small clump into the holes in a dead coral skeleton.

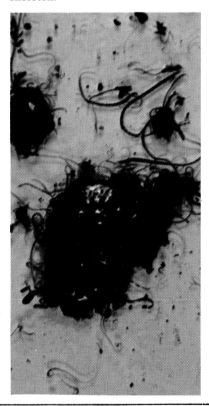

For tubeworms, anemones, clams, and certain other types of sessile filter-feeding invertebrates, a meat baster can be used to pump a cloud of suitable food into the mouth or feeding arms. This ensures that enough food is available to the animal without it having to depend on just what the currents in the tank bring to it.

skeleton of a dead coral so that it may resemble the living polyps. If you feel affluent, start them on live coral if available and then see if you can fool them later. Try a newly opened shellfish such as a mussel or small clam, which is good also for difficult angelfishes. Once they get the idea of feeding in their new surroundings, they will often switch voluntarily to a variety of standard foods, finishing up with competing heartily with the others for much of what is offered.

Lionfishes and anglers are voracious feeders, but they may have to learn to eat dead food. The trick is to feed them with a few live guppies, gambusia, or any spare small fishes available, making sure to keep them hungry over the first few days. Then offer a dead fish on the end of a stick with a needle attached to spear the corpse—making sure that it doesn't protrude and harm the predator. Wave it gently around a little and away from rather than toward the fish, and the fish will very likely follow and seize the offering. If it doesn't, try again later. It is most unusual for more than a week or two to pass before the fish will be taking any piece of food from the end of the stick, and soon it will look for the food as soon as you go up to the tank. The only sad thing about this helpful technique is that anglers lose the habit of playing their bait, so that intriguing display is lost. Remember also that all predators benefit tremendously from at least an occasional feed of whole small fishes, alive or dead, that provide a complete diet for them.

Mention has already been made of the necessity to keep slow feeders and planktonic feeders on their own or with others of the same disposition. There is virtually no alternative to this, as to provide enough to give them a chance to eat sufficiently would involve gross overfeeding, with the one exception that they can be kept with fishes not interested in eating either them or their food. Thus, seahorses, pipefishes, and mandarins can be housed with *small* lionfishes, which will not be able to swallow them and will not try as long as they are well fed. The lionfishes will not be interested in adult brine shrimp or mosquito larva unless they are themselves very small, almost fry, when they do eat such foods. Even so, their feeding habits are such that they cannot eat very fast and a tiny lionfish would leave plenty for the others to eat. Watch out as they grow up though!

Quite a number of young fishes are cleaners, but very few keep the habit into adulthood. The cleaner wrasse *Labroides dimidiatus* is a well-known species that retains cleaner instinct, as does also the Neon Goby, *Gobiosoma oceanops*. The fishes in an aquarium would have to be dreadfully infected with parasites for a single specimen of any cleaner to be able to make a living cleaning them, so it must be fed. A cleaner does much the best with at least an occasional feed of livefood. Young chaetodons and angel fishes are usually good cleaners, but they lose the habit as they grow up and switch over

Perhaps the reason that many butterflyfishes and angel fishes are difficult to keep in the aquarium is that the young often have a different diet than the adults. In some cases the young are cleaners and they do not change diet until after a size at which they are collected for the aquarium. *Pygoplites diacanthus* is one of the more difficult angels to keep.

The Neon Goby, *Gobiosoma oceanops*, is a good example of a fish that is a cleaner in nature but would never survive for long if not fed in the aquarium. Unless the other fishes are over-run with parasitic isopods and crustaceans, cleaner gobies will simply starve in a few days. They must be fed just like any other fish.

Precautions must be taken if cleaner blennies are kept in the miniature reef aquarium. First, be sure there are no mimic blennies such as *Aspidontus taeniatus* (above) in the tank. Next, be sure there are not more cleaners than the other fishes in the tank can tolerate. Lastly, be sure the cleaner is well-fed and not dependent on cleaning to survive.

to feeding on such fare as corals and sponges. This switch in diet from plankton or parasites is common to most young fishes, but chaetodons delay it until they are around a length of 2″ in many cases, one of the reasons why it is difficult to raise them in the aquarium, as they need a lot of livefood available almost constantly to flourish and are apt to resist a switch to other available foods.

Despite the usefulness of cleaners in a fish tank, it is not always wise to keep one if there are only a few other fishes present. The cleaner is likely to become a pest, not waiting for fishes to come along and offer themselves for "treatment" as they do in nature, but pestering them until they get so irritated that they chase it away. Beware also of the blenny *Aspidontus taeniatus*, which looks like a cleaner wrasse and takes advantage of its mimicry to nip scales from fishes instead of cleaning them. It is not a common fish (indeed, it must not be to get away with mimicry successfully), but I have seen it among a batch of *Labroides dimidiatus*. Most mimics seem to take advantage of a close resemblance to a nasty-tasting or dangerous species to avoid being attacked, but *A. taeniatus* does things the other way round and mimics a useful fish to get a meal. Either way, they have to be much fewer in number than the species they mimic or the deception would not work.

(Facing page) To have a truly successful aquarium your tank must be disease-free. No matter how attractive the fishes and background, if the fishes are sick all the time you are not a success.

Some of the diseases of marine fishes are the same as those of freshwater ones, but some are not. The conditions favoring disease are much the same in both cases—overcrowding, poor feeding or poor maintenance, anything leading to lowered resistance, and, of course, the introduction of a disease or parasite because of lack of precautions. Relative to the conditions in the ocean, even the most sparsely populated aquarium is crowded, and the concentration of a disease-causing organism in the water is going to be much greater once it gets a hold. Almost every fish in nature has something wrong with it, often several things, just as we have, but it won't have a great infestation of any particular disease—it would be eaten by some other creature before this occurs. Typically if it has a condition like white spot or velvet there will be only a few parasites present, not enough to matter. This is because the chances of the free-swimming stages finding a host are much smaller in nature than in the aquarium.

Overcrowding is one of the main stresses leading to outbreaks of disease in the aquarium. Marine tanks are much more easily overcrowded than freshwater tanks because of the more complex chemistry.

Almost any fish, even a fine specimen like this *Pseudochromis porphyreus* on the facing page, will have traces of disease conditions and parasites if you look closely enough. Your job is to prevent these from becoming obvious and dangerous.

This mono, *Monodactylus sebae*, is infected with the cysts of *Henneguya*, a myxosporidean protozoan. Such infections are usually terminal and are also contagious. Since there is no effective treatment or cure, it is best to quarantine such a fish until the identification of the parasite is confirmed and then euthanize it.

Immunity

There is an additional factor at work, the fish's own resistance to disease. In the fish tank there is often a sudden burst of infection as hordes of bacteria, fungal spores, white spot tomites (the free-swimming stage), or whatever else are released into the water. The fishes receive a heavy dose immediately and have little or no resistance to it; then before they can build up any resistance they are overcome by the infection and if untreated they die. Perhaps because of previous recent exposure to a particular disease, a fish *does* have some natural resistance; it is then weakened by a rise in ammonia in the water and gets heavily infected despite its partial immunity. Neither of these happenings is likely in nature, although similar ones *can* occur. An example is a "red tide", massive pollution of the water by algae that themselves poison the fishes without waiting for an infection to supervene. Normally the vastness of even a local part of the ocean prevents any changes of a dangerous nature or any great concentration of disease-causing organisms.

These factors can be seen at work in the aquarium, particularly in the marine aquarium, which usually is less densely populated than the freshwater aquarium. If you get an attack of white spot in a freshwater tank, it is usually necessary to treat it or heavy losses are likely. In the marine tank it often happens that the attack dies down if no treatment is given. Waiting is a risky thing to do, but it works in the case of many diseases, although certainly not all. Moreover, the disease is not eliminated. What has happened is that the fishes' resistance has had time to build up before the disease becomes severe and there is then a relatively stable, low degree of infection that may vary from fish to fish. Some will show nothing, others will have a few spots, but only a few, that don't much worry them. Sometimes a rather tender specimen such as a chaetodon will show quite an infestation and then lose it for a period, later to become infected again. Put a new fish into the tank and it may fall victim to the disease and die because its resistance was low. On the other hand, it may suffer moderately for a period and then recover. Often it will depend on whether the newcomer is harassed or whether conditions in the tank exactly suit it.

A state of immunity can be maintained by periodic assaults of a disease, and this is what happens in many a tank—just as it happens to you and me. A fish from foreign parts suddenly subjected to conditions in a home aquarium is in severe danger of falling victim to various infections it has never met before or to strains of ones it has met that differ from those to which it was accustomed. Luckily, this rarely happens, simply because the fish passes through various hands before it reaches us and is often a survivor from a batch of which many died in the tanks of shippers and dealers, where conditions are rarely ideal. Most often, it will have been medicated to suppress diseases rather than to cure them, as full treatment for many a condition takes too long and wastes space as well as time. Obligatory quarantine for a substantial period is gradually being introduced in some countries and may eventually help a great deal.

This salmon is heavily and obviously infected with *Aeromonas liquefaciens*, one of the bacteria most often found in conditions of tail and fin rot. For such an extensive infection to have developed there must have been extreme stress on the fish.

The presence of a single large parasite, such as this gill copepod, *Lernaeocera branchialis*, may not indicate a very stressful situation, as a few parasites are found on almost any fish.

Recognition of Disease

Many diseases that manifest themselves by behavioral changes or external symptoms such as spots or blemishes are fairly easy to recognize, but those attacking internally are harder to diagnose and often remain undetected until severe trouble occurs. There has thus been an emphasis in popular books and articles on the treatment of external conditions, while internal diseases are usually harder to treat as well as harder to detect. Knowledge of the handling of internal diseases comes most often from hatcheries in the case of freshwater fishes, and some of it is applicable also to marines. However, it often requires laboratory determination of the nature of the causative organism as well as how to treat it—all beyond the scope of the home aquarist.

Further problems are those of handling infected fishes. It is often impossible for the aquarist to catch out and quarantine a sick fish. If he has a large well-decorated aquarium it may be necessary to wreck it to get a particular fish out, and even then, how is he to know if the others are free of the infection? He may have no quarantine facilities, even if the fish were to be caught. Yet it may be impractical to use many a recommended treatment in the

Although these *Chaetodon semilarvatus* and the *Heniochus intermedius* are gorgeous specimens and appear healthy, they still should be quarantined before being added to another tank. Many diseases are not apparent for days after the change to a new tank and the correlated increase in stress. One infected fish in a tank can lead to infections of all the other fishes in the aquarium.

tank as it may kill nitrifying bacteria or invertebrates, may stain furnishings, or may color the water so deeply that nothing is visible. There are times when a fish *must* be removed from the aquarium, and times when a deleterious remedy must be used, but let's avoid both if at all possible! For these reasons the recommendations that follow will rarely include the use of substances like methylene blue and various other dyes, potassium permanganate, or general tank dosages with antibiotics. If the latter have to be used they should if possible be given by mouth as with ourselves, injection normally being out of the question.

When a fisheries expert runs up against disease, he will as often as not kill some of the sick fishes and dissect them, take specimens for examination under the microscope, and culture for bacteria or other organisms so as to decide the cause of the disease. You and I will not wish to kill our fishes to decide what is wrong with them—it defeats our purpose, even if we have the equipment and knowledge to examine the specimen. So we must go by external symptoms and hope for the best. Luckily, we shall often be right.

The question is: How do you remove a single fish from a heavily decorated aquarium without demolishing the tank decor? The answer is that often you simply cannot remove a single fish. This is one more reason to have a quarantine tank if you are going to be serious about keeping expensive marine fishes. Remember that most chemicals used to treat diseases and parasites are also toxic to invertebrates (often especially so to crustaceans) and even to the plants.

The ability to diagnose diseases in living fishes is one that is acquired through knowledge of both the literature and the appearance and behavior of the fishes in the aquarium. For instance, this photo shows two species of shrimpfish, the more common *Aeoliscus strigatus* and the rarer *A. punctatus*. At first glance it might appear to the inexperienced aquarist that what we have here is a single species of fish with some individuals infected by black-spot disease, the encysted larvae of a trematode. Treatment of the "infected" *A. punctatus* would probably result in their death.

Anything that is brought into an established tank may bring with it disease organisms. It is extremely difficult to medicate a tank that contains such a variety of different organisms as a reef tank. Therefore, if possible, a quarantine tank should be set up for new arrivals.

Treatment With Filters Present

We must remember that many effective remedies are removed by undergravel filters, carbon filters, and resins, and that these must be turned off as far as is feasible. Carbon and similar filters are easy to manage, as no harm comes of turning them off, but biological filters are a different proposition. Turned off, they die and will pollute the aquarium when turned on again. There is no general agreement about how long they can safely be turned off; some say as little as two hours, others much longer, but they certainly cannot stand days of inaction. Antibiotics are probably the most dangerous drugs to biological filters. Luckily they mostly *suppress* bacterial activity and growth rather than kill the bacteria, so if an antibiotic *must* be used in the tank as a whole (which is rarely), turning the biological filter on for short periods at frequent intervals is the best solution. Heavy doses of such drugs as formalin and copper will kill a biological filter, but in recommended doses they are stated to be fairly harmless, reducing but not severely affecting the filter's activity, although they may themselves be partially removed by the filter.

To meet the demand for effective "cures" for marine organisms a broad spectrum of medicines have been developed specifically for them to supplement those already in use for freshwater aquaria. Photo courtesy of Aquarium Pharmaceuticals, Inc.

Symptoms

The symptoms of disease or parasitism may be conveniently divided into the physical and behavioral changes visible to the aquarist. Sometimes they combine to give an almost certain diagnosis, and often they at least suggest a limited group of possible causes for which treatment may be given. It is quite in order to give more than one treatment; in fact, many commercial preparations do exactly that. If there is any doubt about the suitability of combining treatments, try them together in a glass of aquarium water to see that they do not visibly interact by causing cloudiness, a color change, or a precipitate.

The various symptoms you may observe are listed in Table 4 with the probable cause or causes they suggest. Details of the various diseases are given later, together with suggested treatments. These are not always the same as would be given to freshwater fishes with an identical disease because some drugs act better than others in different salinities, or even because a trusted freshwater treatment has not been used sufficiently with marine fishes for us to be able to recommend it.

Further diagnosis than Table 4 permits is very difficult without examining skin or gill smears or dissecting the fish and its organs, culturing bacteria, etc.

We shall now deal with the various diseases of marine fishes found in the aquarium, at least as far as present knowledge allows.

TABLE 4

Physical Symptoms	Probable Cause(s)
Small white spots on fins or skin	*Cryptocaryon irritans* (white spot)
Small white spots that come and go	*Gyrodactylus* or other flukes
Tiny white spots or velvety appearance	*Amyloodinium ocellatum* (velvet)
Whitish fluffy patches on skin or fins	Probably *Lymphocystis, not* fungus
Gray or white patches on skin or fins	*Chondrococcus columnaris* (bacterial)
Nodules visible beneath the skin, white or dark	*Plistophora, Glugea,* or *Henneguya;* Cestoda (tapeworm larvae)
Nodular white swellings on skin or fins	*Lymphocystis* (a virus)
Red streaks on skin or fins	*Vibrio* or other bacteria
Destruction of tail or fins	*Pseudomonas, Vibrio* (bacterial)
Ulceration of skin	*Ichthyosporidium* (a fungus) *Vibrio* or other bacteria
Yellow to black nodules on or below skin	*Ichthyosporidium*
Wasting, hollow belly, possibly sores	*Mycobacterium marinum* (tuberculosis) or underfeeding
Scale protrusion, often reddish, with normal body	Bacterial infection of scales
Scale protrusion due to an inflated body	Dropsy
Exophthalmos (pop-eye) Cloudy eyes.	Gas embolism, copper poisoning,
Cloudy eyes, even blindness	Toxins, severe white spot or velvet.
Destruction of lateral line	Viral attack?, cause uncertain
Crustaceans on skin	Copepods, *Argulus* etc.
Spinal deformity	Tuberculosis, vitamin or calcium shortage, genetic condition, or *Ichthyosporidium*

Behavioral Symptoms	
Glancing off rocks or coral	Velvet or white spot, toxins
Fins clamped	Velvet or white spot, toxins
Fins constantly erect	A sign of impending death
Loss of balance, sluggishness	Water too cold, possible disease
Severe loss of balance	Swim bladder disease
Gasping at surface	Water too hot, O_2 lack or CO_2 excess, toxins
Sudden dashes, even out of water	Lice or flukes, low pH, toxins
Unusual colors, dark colors	*Ichthyosporidium*, toxins, nervous system disturbances
Failure to eat as usual	Bullying, toxins
Failure to eat from start	Wrong food, cyanide poisoning

They are presented in order of size of the infective or parasitic agent, from viruses to flukes and crustaceans. The remedies recommended follow the principles already outlined. Many are commercially available, but not too often in simple form. There is a tendency to combine drugs, etc., into a single package offered as a cure-all for a whole host of diseases and not necessarily as good as perhaps a higher dose of any single constituent for a particular, identified disease. Also, most commercial remedies are for freshwater fishes and are not always suitable for marines. Chelated copper preparations (a copper salt bound to an organic molecule that releases it slowly or acts in combination with it) can be a risky way of treating marine fishes, as they cannot be monitored for maintenance of a desired concentration in the water and are very hard to remove from the tank. So think carefully before using a stock remedy; it is often best to get your drug store to make up exactly what is recommended here. Do not use any remedy that does not specify contents; at least know what you are putting into the tank.

With any medication, remember that the theoretical gallon capacity of your tank is never achieved. Avoid over-dosing by allowing for the depth of substrate, the air gap at the top, and the volume of coral, rocks, and equipment in the aquarium.

This is particularly important with a substance like copper and in small tanks, where the theoretical capacity is usually much higher than reality. It is a very good idea when first filling a tank to measure exactly how much water has been added. If

that has not been done, calculate by the usual formula of height x length x width in inches divided by 231, but modify it by subtracting from the estimate allowances for the average depth of substrate plus the air gap and then subtract a further 5% for coral, etc.

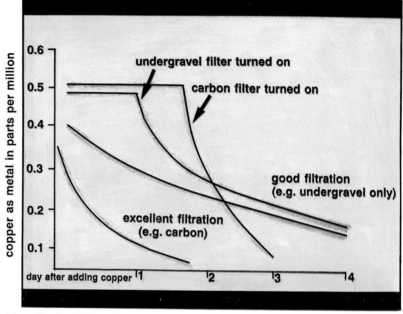

The effects of filtration on copper concentration.

Example:
Theoretical volume of tank = 16 x 30 x 14 / 231 = 29 gals.
Corrected volume = (16—3—1) x 30 x 14 / 231 x 0.95 = 21 gals.
Quite a difference! If you use a divided tank, the correction must be made in two parts, one for the inhabited front section with its substrate and furnishings and another for the back sections, containing practically only water, with no correction for the height and no 5% final subtraction.

Viral Diseases

Viruses are little more than packets of genes in a protective sheath capable of injecting the genes into a living cell, where the virus is capable of reproducing and from which it will spread and infect other cells. They cannot reproduce on their own and are therefore obligatory parasites on the order of 0.0003 mm. (0.3 micrometers) in size. They are suspected of being the cause of cancers and various other swellings and disorders in fishes, but little has been proved except in the case of lymphocystis.

Lymphocystis virus causes host cells to swell up to a tremendous size in what become tumors. These are made of grossly enlarged connective tissue cells and look like a white spawn of some kind on the fins and body and may proceed to "cauliflower"-type swellings in more advanced cases. There is no known certain cure, although there are one or two preparations on the market that claim to cure it. The disease is highly infectious but does not often kill its victims, only disfiguring them. An infected fish is best destroyed.

However, in a well-maintained aquarium, the disease can disappear leaving little or no

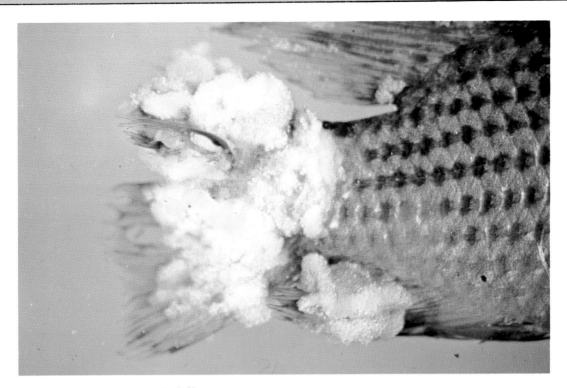

The most common viral disease
that can be recognized among
aquarium fishes is lymphocystis, a
condition characterized by
cauliflower-like tumors such as
these on the caudal peduncle of
the fish above.

In many marine fishes
lymphocystis never develops into
very large tumors (below), but the
disease is just as incurable and
contagious as if the cauliflower-
like lesions were present.

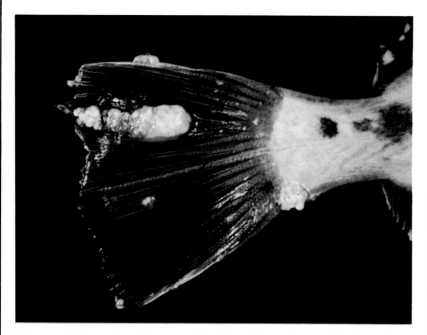

Close-up of the cysts or tumors of lymphocystis on the caudal fin and peduncle of a flounder. Their somewhat granular nature can be seen here.

Bacterial Diseases

Bacteria are single-celled organisms of a higher order than viruses that usually can grow and multiply without parasitism. They are still minute but are much bigger than viruses and of variable shapes and sizes. They are also susceptible to treatment. Very many different species are responsible for diseases in fishes as in other creatures, and there will usually be no opportunity to establish exactly which species is the causative agent. While indications will be given as to which bacteria are probably responsible for a particular condition, the problem will be tackled in general terms under the various recognizable symptoms. Even normally harmless bacteria may cause an infection if aquarium conditions are bad enough.

scarring. A really precious fish may be worth keeping in the hope of this happening, preferably in quarantine.

Although a number of viral diseases causing such conditions as septicemia and pancreatitis are known or suspected in freshwater fishes, no such conditions are known to occur as a result of viruses in marine fishes.

Lateral line disease is a condition in which the lateral line system is attacked by what appears to be a virus, although this is not certain. The disease usually starts at the head end of the system, giving an appearance as if parts of the lateral line have been gouged out, leaving a groove in the tissues. This may later spread over the greater part or even all of the lateral line. The condition may affect only one fish, or only one species, and the observations of some aquarists suggest that it may be nutritional in origin or due to poor husbandry, but the finding is not

general. There may well be more than one disease involved, and there is clearly a lot more to be learned about this condition. It is probably best to get rid of an infected fish or, if treatment is decided upon, to remove it for individual attention in an isolation tank. Sometimes a change of environment effects a miraculous cure with cases of this type.

Red streaks on the body, fins, or tail that may proceed to ulceration and to parts of the fins and tail dropping off, then designated as *fin rot* or *tail rot*, are caused by many different

Lateral line disease is especially prevalent—or at least most easily seen—in surgeonfishes. Often the canals of both the head and the body are affected, resulting in a most unsightly appearance. A virus associated with the condition is shown in the insert. Such fish usually are destroyed.

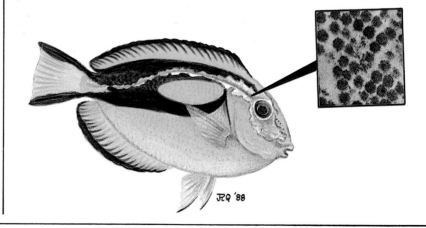

bacteria. The most common would appear to be species of *Pasteurella*, *Vibrio*, and *Pseudomonas* (*Aeromonas*), all Gram-negative bacteria that are insensitive to penicillin. External treatment is only effective if the infection is very light or if high doses of antibiotics are given— which we try to avoid. If the infection is caught in the early stages, you can try treating the whole tank with a disinfectant, cleaning the tank up as far as possible, and feeding lightly for a few days. A good disinfectant is acriflavine (trypaflavine) or monacrin (mono-amino-acridine). Make up either as a 0.2% solution and give up to 1 teaspoon (5 ml) per gallon. Acriflavine gives a yellow tinge to the water and monacrin an attractive bluish ocean-like appearance. Monacrin was even added to the water by some friends when photographing fishes to give a more convincing picture, so you can guess that it is a very unobjectionable cure. Further treatments can be added every few days as the colors subside, an indication of bio-degradation of the drugs. If there is no sure cure in a few days, however, do not persist, but switch to the next treatment.

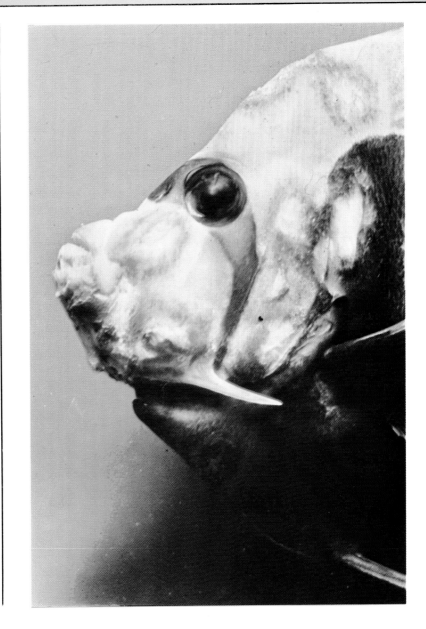

(Above) A bacterial infection, possibly *Vibrio*, of the head of an angel fish. The surface tissue is largely destroyed. Such a fish would be very subject to other infections.

(Left) A large lesion infected with *Aeromonas*. This will probably spread and may lead to death.

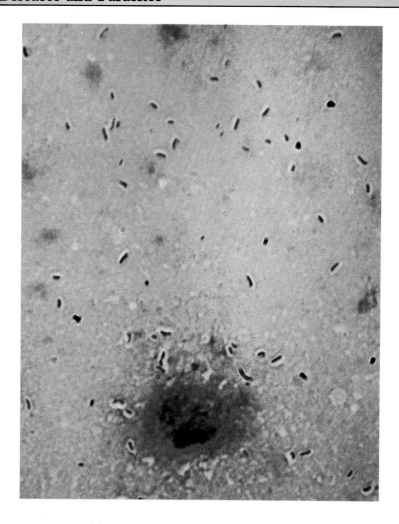

and 1 oz is sufficient for a lot of feeding, about a thousand fish-doses on average. If dry food is not suitable, the antibiotic can be mixed with chopped frozen or fresh foods, even with tubificid worms. If the fishes affected will not eat, there are two alternatives. Either catch the ones affected, if only a few, and treat them in an isolation tank with antibiotic in the water, or if this is not practical, treat the whole tank. Avoid doing this if you possibly can, as there then arise the problems of effects on filters or of filters on the drugs themselves. But if you are reduced to general dosage in the water, give effective amounts often enough to expect a cure. These are *at least* 50 mg per gallon of any of the drugs mentioned except neomycin, when 200 mg per gallon should be given. Repeat these doses at two-day intervals until a cure is achieved or the treatment is clearly ineffective, in which case switch to another drug. As you will be aerating briskly since filters will be turned down or off, it is best not to use aureomycin because it froths badly and also turns red in sea water.

Tuberculosis in fishes is caused by a specific bacterium, *Mycobacterium marinum*. Affected fishes waste away, with a hollow belly, affected kidneys, blotched skin and various other symptoms in the advanced stages. Earlier they show listlessness, poor appetite, and perhaps skin degeneration and ragged fins, when tuberculosis must be suspected. The disease is highly infectious and difficult to cure, so that any isolated case should preferably be removed and destroyed unless very precious. If you decide to treat the disease, either in a quarantine tank or in the main aquarium itself, the

This condition is or will become a bacteremia, an infection of the blood as well as tissues, and is preferably treated from within. This is done by mixing a suitable antibiotic with the food in a concentration of around 1% by weight. Chloromycetin (chloramphenicol), aureomycin, neomycin, and gentamycin are all likely to be effective, with preference for chloromycetin because it has a very wide spectrum of effectiveness and is rarely used in human medicine. So if we do create a chloromycetin-resistant strain of bugs, we are unlikely to cause trouble to our own species. Chloromycetin tastes horribly bitter to us, but fishes don't seem to notice anything. The best food with which to mix the drug is a

Identification of bacteria usually is dependent on complex chemical tests as well as the detailed appearance of colonies on various culture media. Even when examined micoscopically, most bacteria are rather featureless.

dry flake, taking care to get an even mix. As long as the fishes are eating, and if they are kept a bit hungry, they will snap up the food plus drug before much of the latter dissolves in the water. Twice daily feeding is enough to keep up a good blood level of antibiotic.

Antibiotics usually come in capsules of 250 mg, sometimes 500 mg, so make sure which you have. A 250 mg capsule added to 1 oz of dry food is just under 1%, same drugs as are used for human and bovine tuberculosis may be effective in fishes, as the causative bacterium is a *Mycobacterium* in these species as well. The drug of choice is isoniazid, which can be fed to the fishes as recommended for the antibiotics and by the same techniques. Alternatively, it may be given in the water, since the fishes may be eating poorly, at 50 mg per gallon repeated every third day, with siphoning off of 25% of the water in the tank. Treatment may take up to 2 months.

Dulin recommends that rifampin be given in addition to isoniazid to hasten a cure. Rifampin must be fed, as it is not soluble in water, and can be given as for an antibiotic, but at a lower concentration, about 0.01%. He gives a formula for making up a suitable food containing rifampin in a liver-pablum mixture (Dulin, *Diseases of Marine Aquarium Fishes*, T.F.H. Publications, 1976). Writers on freshwater fishes do not recommend such a treatment for tuberculosis, although it is also caused by a *Mycobacterium*, but not *M. marinum*. Why this is so is not clear.

Always remember that tuberculosis is a sign of poor conditions and that well-kept fishes will not develop it unless overwhelmed by an infection from tankmates. With care not to introduce suspicious looking specimens, your fishes should never be troubled by it. It is, by the way, possible to acquire a local infection yourself if you plunge hands that are cut or abraded into a tank containing tuberculous fishes. Luckily, the infection does not spread and is not serious, but it is just as well to avoid it.

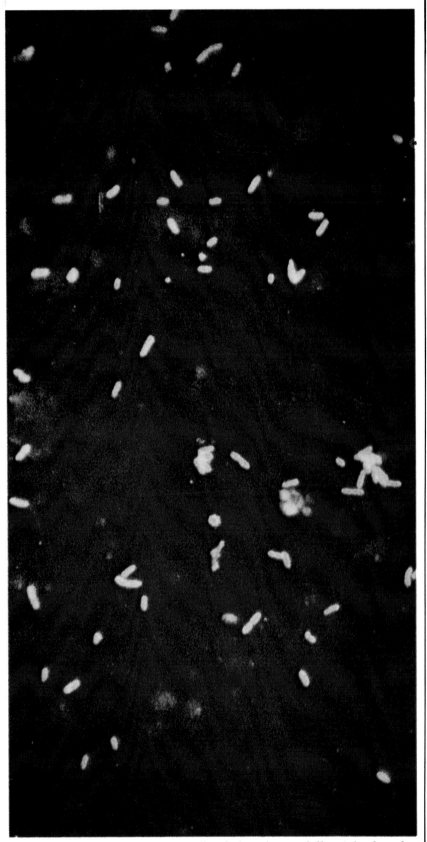

When the bacterium *Aeromonas liquefaciens* is specially stained and treated, it appears bright green under a fluorescent microscope, one of the distinguishing features of the species.

Dropsy is a swelling of the abdominal cavity, also called ascites, and is usually due to kidney disease. Fluid that should be eliminated accumulates in the abdominal cavity and causes the fish to swell up and its scales to protrude. Sometimes pop-eye (exophthalmus) develops as well. The causative organism is normally a *Corynebacterium* species in aquaria, although the condition can be caused by other bacteria in nature. The curious thing about *Corynebacteria* is that they are the only Gram-positive ones known to cause disease in the aquarium. Gram-positive bacteria are sensitive to penicillin and erythromycin, the latter

being the antibiotic of choice in the aquarium. Give it by mouth exactly as for other antibiotics at a concentration of 1%, or if necessary into the tank at 50 mg per gallon, repeated every two days with 25% of the water removed each time.

Be careful to distinguish true dropsy from scale protrusion, which may be caused by other bacteria. With scale protrusion there is no actual swelling of the abdomen, just reddened scales sticking out from the body. A common cause is *Pseudomonas* infection. The condition should be treated just as for red streaks or fin and tail rot—the antibiotic chloromycetin by mouth for preference.

Chondrococcus columnaris the cause of dreaded "mouth fungus" in freshwater fishes, sometimes infects marines. However, it does not take the same form, but causes gray-white lesions of the skin and fins and may infect the gills. Treat with chloromycetin by mouth as for red streaks etc.

Protozoan Diseases

Protozoans are single-celled organisms, but they can be large enough to be seen by the naked eye, if only just. There are various classificatory systems for these animals, but I have adopted a simple although rather antiquated one that suits our purpose in describing classes of protozoans infecting or parasitizing fishes. The Flagellata have one or more whip-like processes with the aid of which they swim around and may attach themselves to fishes. The Sporozoa are all parasitic and form spores, hard resistant capsules in which young forms develop or the adult resists adverse conditions. The Ciliata have a surface covered by numerous hairs, the cilia, by means of which they swim and settle onto their hosts.

Flagellata

Velvet or coral fish disease is very common in aquaria and must be suspected whenever a purchase is made, yet it can be hard to see. It is primarily a gill infection, and by the time it becomes visible externally there may be heavy infestation. It shows up on the body and fins as a whitish powdery dusting over the surface, easiest to see on dark surfaces and when the fish is facing you and obliquely inclined to the light. Look for it especially on anemonefishes as they are particularly prone to the disease

Dropsy usually is caused by bacteria of the genus Corynebacteria, which respond to penicillin and related drugs. Perhaps no two cases of dropsy are exactly alike, but they usually involve some type of kidney damage or malfunctioning that causes fluid to accumulate in the body cavity. Often the eyes protrude, the familiar "pop-eye" condition that may also have several other causes. Some types of dropsy do not respond to medication and are probably caused by other types of bacteria.

yet often show little or no discomfort. If they are breathing rapidly, suspect velvet disease that has not yet shown up externally.

The cause of velvet is *Amyloodinium ocellatum*, related to the freshwater *Oodinium* species. It is a dinoflagellate about 0.01 mm in diameter, with a long and a short flagellum. It settles onto the fish's gills or skin and adheres at first by its long flagellum, then puts out finger-like pseudopodia (false feet) that invade the tissues and give it a good grip. These also enable it to feed on the host, eventually forming a cyst embedded in the skin. This cyst drops off after a few days or may in some cases remain embedded on the fish, but in either case it gives rise to about 200 new dinospores, the free-swimming stage. These of course swim around and settle on a new host if available, the whole cycle taking an average of ten days in the tropical marine tank, the actual time depending on temperature and perhaps the strain of the disease. There may even be other species of *Amyloodinium* with slightly different life histories, as accounts of the process differ with different observers.

The symptoms of velvet, other than the powdery appearance and rapid respiration already noted, are glancing against objects in the tank, clamped fins, and a peculiar wobbling motion like swimming without getting anywhere (often seen in perfectly healthy anemonefishes, however, but not in other species). Secondary infections may follow a heavy infestation, causing bloody streaks and ragged fins and tail and eventually death. Some of the earlier symptoms are seen with toxic water, but if a clean-up and partial change of water do no good, go on

Schematic life history of the flagellate *Amyloodinium,* the cause of marine velvet or coral fish disease, a common condition among captive fishes. Because the treatment uses copper compounds, any plants and invertebrates in the tank are at risk.

suspecting velvet even if you cannot yet see it. If trouble persists, with many fish affected, consider treating for velvet.

Treatment of velvet is best with copper as the citrate, which is more soluble in sea water, or the sulphate, which is easiest to obtain and perfectly satisfactory. Fishes are susceptible to copper poisoning, so care over dosage is necessary, but luckily marine fishes tolerate it better than do freshwater ones. Copper is also toxic to algae, so if there is a heavy coating over rocks or coral, most should be removed before treating a tank with the metal. It acts in two ways, killing the dinospores and causing the fishes to produce a copious mucus that makes it harder for any surviving parasites to attach themselves. It may also help the shedding of cysts, as they tend to disappear

from the fishes quite rapidly under copper treatment even when not all could have been ready to become detached.

To treat with copper, it is necessary to maintain a concentration of metallic copper at not less than 0.15 ppm for not less than 10 days, during which time all the dinospores that are going to hatch out will do so and be killed by the metal. As copper is so toxic, we do not wish to exceed, say, 0.20 ppm, and the actual concentration should if possible be monitored by daily readings. Much of the copper will be removed by corals and the substrate, even if filters are turned off, so additional doses will be needed as the concentration falls, the amounts depending on how much it falls. Copper test kits are not all that reliable, but they are the best we

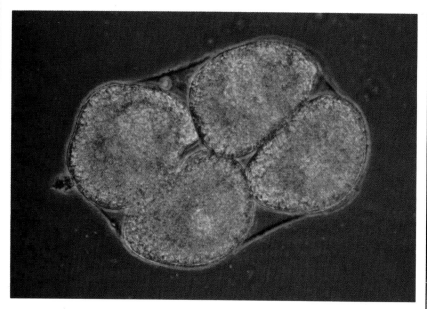

Close-up of the cells of *Amyloodinium ocellatum*, the cause of marine velvet. This encysted stage is very hard to destroy.

can do and better than nothing. If no kit is used, add half doses at two-day intervals. Most marines can stand up to 0.4 ppm, so we actually have a fair margin of safety.

The ordinary blue crystals of copper sulphate are $CuSO_4 \cdot 5H_2O$, containing only one fifth their weight of metallic copper. To give a dose of 0.15 ppm, make up a 1% solution of the crystals (10 g per liter or 38 g per gallon) in distilled water, as tap water may result in a precipitate eventually if not immediately. Then add 0.28 ml per gallon to the tank, or *one standard 5 ml teaspoon* per 18 gallons. Do not use chelated copper preparations as often recommended; they cannot be monitored and do not give as reliable a cure as the simple salt. The signs of copper poisoning in fishes are gasping, turning onto the sides, and exophthalmus (pop-eye). If any signs appear, remove the copper by turning on the carbon filter or by adding an ion exchange resin such as Zeolite 225 to any filter.

Copper is mildly toxic to vertebrates in large concentrations, but it is deadly to invertebrates in even trace amounts. Never treat a fish with copper compounds in the invertebrate tank.

Copper cannot be used in the presence of invertebrates or precious growths of algae. A suggested alternative of Spotte's is quinine (used for freshwater white spot), which also has the advantage of not killing bacteria. I have no experience of quinine for *Amyloodinium*, but the recommended dosage is 2 g quinine hydrochloride, or quinine sulphate if the former is unobtainable, per 100 liters of water. This is 77 mg per gallon. It is also recommended to change 50% of the water and add a half dose of quinine on the third day. The minimum treatment period is five days. This seems very short to me, and I would continue for ten days, just in case, as the dosage suggested is quite low and harmless to the fishes.

Sporozoa

Coccidia infestation is not uncommon in shoal fishes like the herring but rarely affects aquarium fishes. Those prone to it seem to be eels, where the intestinal epithelium is attacked. However coccidia, of which there are many genera and species, may

infect almost any organ. The disease is essentially internal and rarely recognized. A cure has not been described.

Myxosporidia are common causes of fish disease, of which *Henneguya* species have been recognized in the aquarium. They cause large cysts almost anywhere in or on fishes. When visible externally they are known as tapioca disease, from their whitish appearance. These can be confused with *Lymphocystis*, but this is of no importance, since in common with it no cure is available and infected fishes are best destroyed.

Microsporidia are frequent parasites of the muscles of fishes. *Plistophora*, of which many species are known, is responsible for neon tetra disease in freshwater fishes. In marines it may cause large cysts forming visible bulges in the tissues and believed to form spores that further infect the host as well as being released into the water to infect other fishes, as with *P. hyphessobryconis*, the cause of the freshwater disease. Another microsporidian is *Glugea*, the most commonly recognized species being *G. anomala*. This also causes large intramuscular cysts and has been reported as causing mortality in marine sticklebacks. Different *Glugea* species infect many food fishes and may attack the sex organs and skin, causing serious losses to fisheries from reduced fertility, poor growth, and even death, as well as looking repulsive to the customer. As with the Myxosporidia, no cures are known for any of these conditions, so the tale for the sporozoans is indeed a gloomy one.

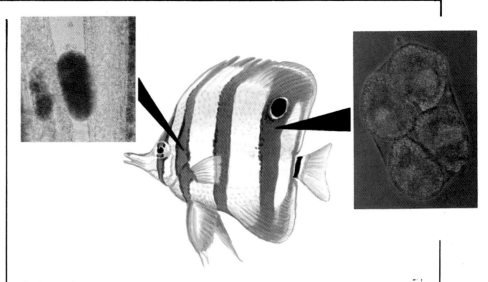

A chaetodon infected with both white spot (*Cryptocaryon irritans*) and velvet (*Amyloodinium ocellatum*). The cells of the *Cryptocaryon* are shown in the insert and are lodged in the gill filaments. The velvet is encysted in the skin.

A specimen of the triggerfish *Sufflamen bursa* with a large tumor or cyst on the base of the dorsal fin. Often the cause of such a growth is never determined, as it may be due to anything from lymphocystis to microsporids. It is best to consider such fishes undesirable and incurable.

The microsporidean *Glugea* exacts a high toll among many smaller fishes. This stickleback is heavily infected, with obvious cysts on the posterior part of the body.

Cross section through part of a stickleback infected with *Glugea* (the large round dark object) and the tapeworm *Schistocephalus* (the elongated segmented object).

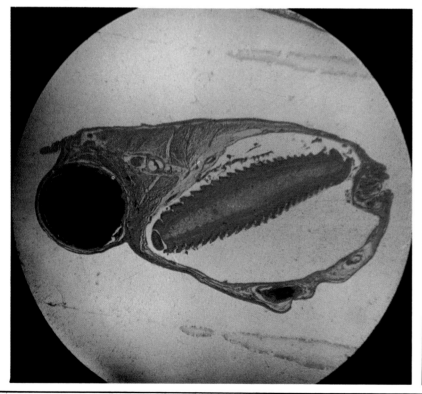

Ciliata

White spot. The only ciliate known to be of importance in marine aquaria is *Cryptocaryon irritans,* the cause of white spot. Like velvet it is a gill and skin infection, but with larger and usually fewer white spots visible. Also like velvet, it is primarily an epidemic disease of aquarium fishes and is usually only found as a light infection in the wild. The spots are not only larger but are deeper-seated than those of velvet and do not disappear very rapidly when treatment is applied. The free-swimming stage, or *tomite,* is the infective agent, a ciliated cell about 0.05 mm long that attaches to the host and penetrates the gill or skin, feeding on the tissues as it digs in as a *trophont,* very much as does freshwater white spot, or ich, *Ichthyophthirius multifiliis.* Cysts (*tomonts*) are then formed that may be up to 1 mm in diameter. These eventually drop from the fish (according to some authors they may stay, trapped in the mucus) and produce up to 200 new tomites over about eight days, that are freed to infect other fishes. It is alleged that they must find a host within 24 hours or die. A fair degree of natural immunity to *Cryptocaryon* seems to exist and it is less of a menace than *Amyloodinium* in many aquaria, although still very much to be reckoned with if it takes hold in a crowded and perhaps badly maintained tank.

The standard treatment for marine white spot is the same as for velvet—copper. As an alternative, quinine may also be used, again as for velvet. If you are quite sure that the disease is indeed white spot and *not* velvet, sodium sulfathiazole or even sulfathiazole itself may be used in the presence of invertebrates. The dosage is one level standard

This young angel fish, *Pomacanthus semicirculatus*, is heavily infected with saltwater ich or white spot, *Cryptocaryon irritans*. Such heavy infections are usually highly contagious and often fatal.

teaspoon per 5 gallons, first dissolved in a glass of fresh water and mixed thoroughly into the aquarium. This may be repeated after a few days if necessary. The treatment is reported as very effective by some aquarists, but I would always prefer copper if feasible. However, some authors find copper less effective against white spot than velvet, and it seems possible that different strains exist with different susceptibilities. Formalin is also added by some and is in some commercial preparations, but it should not be added to the aquarium as it kills bacteria and is far too likely to damage fishes in the long term.

FUNGI

Fungi are plants lacking the characteristic green pigment, chlorophyll, by means of which other plants manufacture sugars, etc., from carbon dioxide and water. Their variety is almost endless and so are their life histories, but rather little is known about their disease-causing capacities in fishes, which in marines appear to be very limited. Although I have myself listed it in the past as a "rather rare disease" of marine fishes, on the evidence of others, I have never seen a case of real ordinary "fungus" in a marine fish nor have others I have consulted. This includes a tuft or tufts of external hyphae (fungal threads) looking rather like cotton wool on the surface of the fish or protruding from the gills, or even a suggestion of it on the body surface. Yet Dulin and Moe both describe *Saprolegnia*, a common type of fungus in

freshwater fishes, as invading marines. Since it is not listed by other recent authors and is in fact *cured* in freshwater fishes by the time-honored salt treatment, we are left to wonder. According to Johnson & Sparrow, cited by Sindermann (*Diseases of Marine Fishes*, T.F.H., 1966), the only fungal disease of note in marine fishes is *Ichthyosporidium hoferi*. The same author notes that *Saprolegnia* infections in Pacific salmon disappeared when the fishes were placed in sea water. There is, however, a rather fungus-like stage of *Lymphocystis* that could easily be mistaken for *Saprolegnia,* and I have seen fishes that on one part of the body looked exactly as if they were attacked by external fungus, yet around the mouth were the cauliflower outgrowths typical of *Lymphocystis* at a more advanced stage.

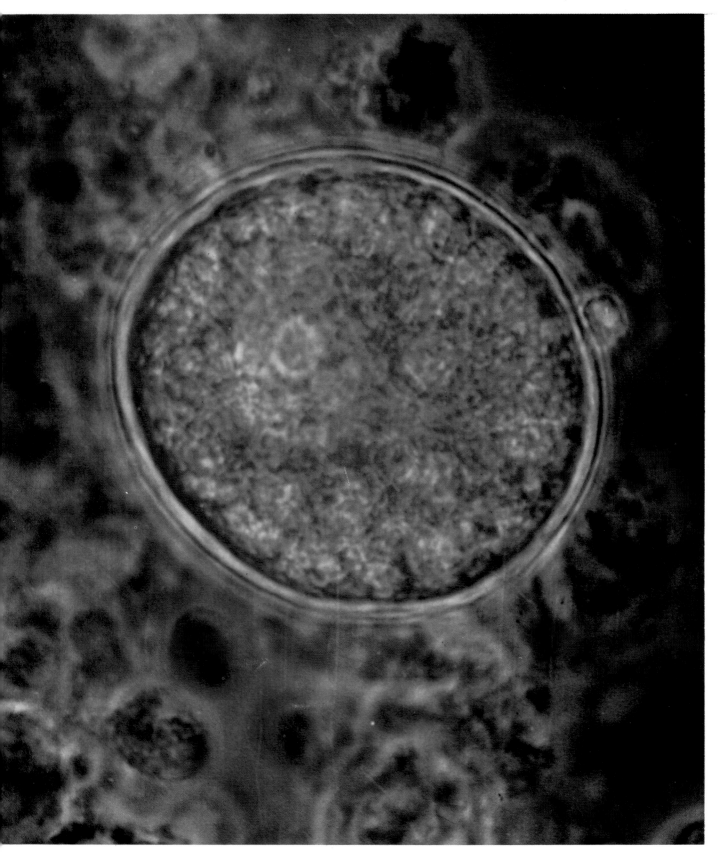

Detail of an encysted cell of *Ichthyosporidium hoferi*. Often the host forms a capsule of connective tissue around the cyst, preventing spread of the daughter cells and thus limiting the disease.

Ichthyosporidium (formerly *Ichthyophonus*) *hoferi* is in itself a source of confusion, because *Ichthyosporidium* is a protozoan and *Ichthyophonus* a fungus. Both names have been used to designate the fungus, and most of the recent literature uses *Ichthyosporidium*, so I shall do the same. Anyway, I mean the fungus! It is a common disease of both freshwater and marine aquarium fishes, first attacking the liver and kidneys but later spreading all over the body. The fungus is eaten with the food or feces of other fishes, invades the bloodstream via the gut, and settles down in the liver, from where it soon appears all over the place. Cysts of up to 2.5 mm are formed; these are brown in color because melanophores accumulate in them. Daughter cysts develop either by budding from the mother cyst or by forming inside it. If inside it, they are freed by bursting of the mother cyst. *Ichthyosporidium* is common in the wild and a cause of fisheries losses, often limiting the catch of herring or mackerel.

The symptoms of infection are dependent upon the stage of the disease and its particular course in individual fishes. Sluggishness, loss of balance, and hollow belly are frequent early signs, often to be followed by visible external cysts, brown or yellow in color, that may break into sores or ulcers and appear almost anywhere. Typically, just a few brownish knobs on the body will be the first sure sign that *Ichthyosporidium* is at fault. The disease is not a rapid killer, and sometimes a fish will limit its spread by its own natural resistance, encapsulating the diseased areas. However, it is never safe to allow an infected fish to stay in an aquarium, and it should be removed in the hope that it has not yet passed on the

This scat has a fungal infection. Since salt usually kills external fungi, increasing the salinity for the fish might effect a cure; the fish had been maintained in brackish water, not full-strength salt water.

disease to others. There has been a good deal of discussion about whether there is more than one species of *Ichthyosporidium*. The infection in marines has been called *I. gasterophilum*, but the experts seem to prefer to recognize only one species, *I. hoferi*.

Whichever is the case, the infection is hard to deal with in all fishes. No definite cure is available, but the treatments recommended for freshwater fishes might well help with marines. These are phenoxethol added to the food as a 1% solution or chloromycetin at 1% in the food also, but I would only attempt to save a very precious fish in quarantine. Early in the infective process, chloromycetin is said to attack the fungus before ingestion or even in the gut, so treatment of a tank from which an infected fish has been removed is a good idea, but by feeding as above, not overall treatment of the water. It is also stated that

250 mg of streptomycin plus 250 mg of penicillin per gallon can cure the disease, if you can afford to try the treatment. Be sceptical about all of these recommendations—antibiotics are not characteristically active against fungi.

Worms

The term "worm" covers a multitude of creatures, once defined as being "any elongated creeping thing that is not obviously something else". Parasitic worms are classified into trematodes (flukes), cestodes (tapeworms), nematodes (roundworms), and acanthocephalans (spiny-headed worms), with a few others for luck. It is often the larval forms that cause the most trouble, encysting in muscle or one or more of the internal organs, whereas the adults more frequently live either freely or in or on their hosts.

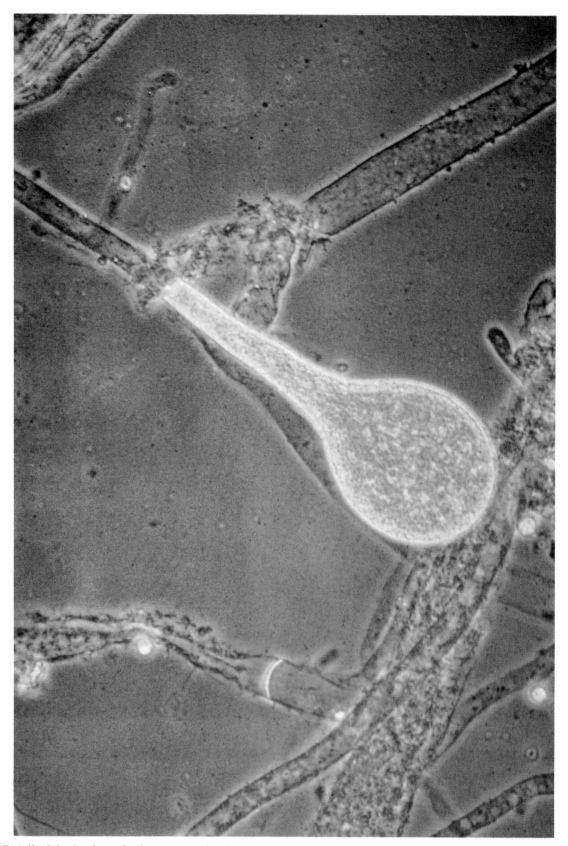

Detail of the hyphae of a fungus erupting from a spore.

Benedenia melleni a parasitic trematode, is one of the worst of the fish flukes, partly because it is monogenetic. This means that it passes directly from one member to another of the same species, whereas many of the flukes need two or three hosts of quite different species to complete their life cycle. The so-called digenetic flukes, needing at least one intermediate host, may pass from a fish to a bird or mammal that eats their first host, thence via its feces their larvae pass to perhaps a water snail, then back to a fish. Naturally, such a complicated chain is not available to aquarium fishes, and except for eggs or larvae introduced by snails or other pond creatures, the digenetic flukes are not a problem.

Benedenia parasitizes externally, settling on the skin, eyes, gills, and elsewhere. It is transparent and attains a length of only about ⅛″, so may be unsuspected when first introduced to an aquarium, but it can multiply to epidemic proportions. Eggs are laid on the surface of the host and take about a week to hatch into ciliated larvae that swim around for only a few hours. If they encounter a fish, practically any fish, they latch onto it, lose their cilia, and feed much as does velvet or white spot on the mucus and skin of the host. (If no host is found, they sink to the bottom and eventually die.) In bad cases scales are lost and open wounds occur that readily become infected by other organisms, notably bacteria. The host scratches itself on objects in the aquarium as if infected by perhaps white spot, for which the flukes must not be mistaken.

Treatment has to be rather heroic, and involves removing the fishes from the aquarium, always a nuisance and sometimes almost a tragedy if you have a beautifully arranged, large tank that has virtually to be torn apart before the fishes can be caught. However, there is no guaranteed alternative. Once the fishes are removed, there are several possible treatments. The least drastic is a formalin bath for one hour in salt water containing one standard teaspoon (5 ml) of concentrated formalin per 5 gallons. If the fishes have been removed as gently as possible so as not to dislodge the flukes, a single bath may suffice, but if flukes remain in the tank they will reinfect the fishes, which will then have to be treated again. The same is true of an alternative, a freshwater bath for up to 10 or even 15 minutes. The majority of fishes can stand this period in fresh water at the same temperature and approximately the same pH as the aquarium, but each fish must be watched carefully for signs of shock, when

it must be returned to salt water. Don't return it to the tank, but to an emergency vessel where it can recover and be inspected for any remaining flukes.

If the prospect of removing fishes from the aquarium appalls you, as it would me in most instances, there are two other possibilities. With a light infestation, put in a parasite picker and hope for the best. I always try to keep a cleaner wrasse (*Labroides dimidiatus*) in any tank; in fact two are best, as they can clean each other. With a cleaner present, flukes are often nipped very literally in the bud. The other possibility which I have never used myself but one recommended by several experts, is to use a pesticide, some of which are effective against both external worms and crustaceans. Unfortunately they are toxic to fishes as well. One has to be chosen that is more toxic to the parasites than to the host, and the margin is usually small. Those recommended are Dibrom, Dipterex and Dylox (Trichlorfon) at 0.25 ppm (1 mg

Cleaner wrasses feed heavily on flukes such as Benedenia.

per gallon approx.), all of which are preparations of an organophosphate insecticide, DTHP, short for dimethyltrichlorohydroxyethyl phosphonate. A more potent agent, Lindane, is also recommended at 0.01 ppm (1 mg per 25 gallons approx.). All are to be added directly to the water, but not by just sprinkling them in as a fish may well snap up a toxic dose. Instead, stir the appropriate dose into a gallon or two of aquarium water and introduce it gradually to the tank. Carbon filters should be turned off, but biological filters are said not to be affected.

There are over a thousand other trematode species known to infest fishes, but mostly as larvae or in the gut. However, a number parasitic on the surface of fishes always of the monogenetic variety, become pests in the aquarium from time to time. Various species of *Gyrodactylus* have been reported—small skin or gill parasites only about ½″ long, easily mistaken at first for white spot. Some desert the fishes at certain times, and any "white spot" that comes and goes puzzlingly must be suspected to be a fluke. Others, unlike *Benedenia*, are species-specific, particularly of the Pacific angelfishes. This is in contrast to the Atlantic angels, whose flukes seem to be non-selective. I had an instance of two angels and one *Dascyllus* affected with flukes among an otherwise fluke-free population in a large tank and removed them to another tank for treatment. However, left overnight they lost their parasites completely and were thereafter perfectly O.K. Unfortunately this is not a reliable method to use! All flukes are best treated as recommended above.

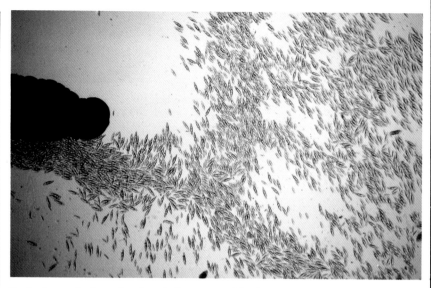
Posterior end of a spiny-headed worm and its just laid eggs.

Cestodes and nematodes are common in the gut of fishes but usually are not apparent unless they kill the host or cause serious wasting, when a post-mortem will reveal them. Sometimes roundworms may be seen around the anus of a fish, when treatment should be instituted. The food should if possible have piperazine mixed into it at the rate of 0.25% (25 mg in 10 g of food) and treatment continued for 10 days. Tapeworms, if for some reason suspected (usually only after a post-mortem of another fish), can be similarly treated with Yomesan, a commercial worming preparation, at the rate of 0.5% (50 mg in 10 g of food), in that case withholding food for a day and then giving a single dose of Yomesan. Dulin (*Diseases of Marine Aquarium Fishes*, T.F.H., 1976), who recommends these treatments, warns that although piperazine and Yomesan are relatively non-toxic to fishes, a massive die-off of the parasites can cause intestinal blockage and the death of the host.

Digenetic flukes do not infest fishes externally, and there are no known cures for them. They occur in many wild fishes, and the only way to try to avoid them in the aquarium is careful inspection of any proposed purchase. Larval forms of trematodes (as well as cestodes and nematodes) are found in the muscles and internal organs of marine fishes and are frequently responsible for commercial losses because of the unpalatability or unpleasant appearance of their flesh. The occasional encysted larva is no problem and is very common—it is only heavy infestations that are going to be noticed by the aquarist. Even then, there is no danger to the other fishes in the aquarium, as the intermediate host or hosts won't normally be present. I say "normally", because some digenetic helminths are passed from one fish species to another, but this would require a predator-prey relationship, so perhaps there *is* a species that lives in the prey of the lionfish, for example, and could thus infect the predator in the aquarium. In nature, a similar

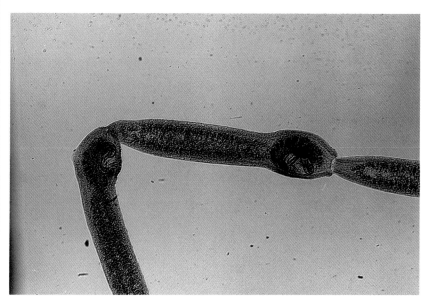

Flukes taken from a freshwater stingray, *Potamotrygon*. Hundreds of flukes occur in marine and freshwater fishes, and their study is both complex and essential.

Close-up of the front end of *Echinorhynchus truttae*, a spiny-headed worm.

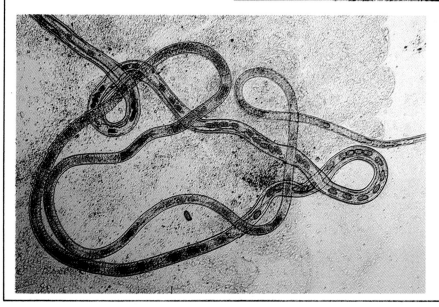

Capillaria is a very common and often destructive roundworm found in a great number of fishes.

A pregnant female oxyurid worm showing the characteristic pointed tail.

example is *Otobothrium crenacolle*, a tapeworm that lives in the intestine of the hammerhead shark. When its eggs are shed they are contained in swimming segments of the worm that are snapped up by other fishes. In them the larvae encyst in the muscles, and their host may later be eaten by another shark, completing the cycle.

Detail of an encapsulated metacercaria (fluke larva) in a gill filament.

Crustaceans

The main aquarium crustacean pests are *Argulus*, the fish louse, of which there are many species, and various parasitic copepods, of which there are very many species found in the wild. *Argulus* species are mostly fairly large and easily seen; they should be spotted before a fish is placed in the aquarium. A typical one would be up to ¼″ in length, with a flattened body and two claws springing from the bases of the first pair of antennae. The mouth is between a series of sucking

discs and is also capable of injecting a toxin that can be fatal to small fishes. The parasites may be quite active, moving about on the host and deserting it for another one. Seahorses are favorite hosts, but any fish may be attacked. Removal is easy with a pair of forceps. Luckily seahorses are easy to catch if you happen to miss seeing *Argulus* before putting them into a tank. Eggs may be laid that hatch out into larvae that then pass through seven molts before becoming parasitic adults. As up to 250 eggs may be deposited by a single female, it pays to get rid of *Argulus* as soon as it is noticed. If eggs are deposited, usually on some object in the tank, the egg masses are visible prior to hatching and should naturally be removed. *Argulus* is not easy to kill by chemical treatments, which should be avoided.

Other copepods can be very dangerous to marine fishes, although many species are quite harmless and may be a source of food. Parasitic species, although numerous and often found in the wild, are not a common cause of trouble in the aquarium. They have various life cycles and can only be dealt with in general terms. Most are quite tiny, but some from large ocean fishes reach a length of 8″. Some penetrate the body of the host and may even invade organs like the heart. *Lernaeocera branchialis* parasitizes the gills of cod and similar species and again may penetrate as far as the heart and great blood vessels. In the aquarium, the problem is confined to the smaller species that infest the skin and gills primarily, but some may invade deeper tissues. The visible signs of infestation may only be egg-sacs hanging from the fish, the rest of the copepod being under the skin, or the animals may be

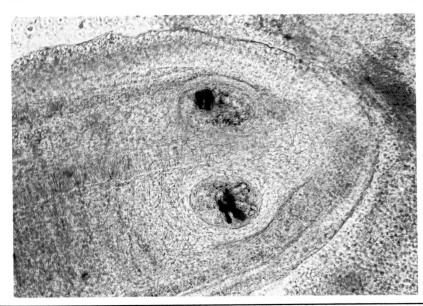

seen hanging from the gills or body. The egg-sacs may look more like worms than part of a copepod and may hang in loops or whorls from the body of the host. Lesions where they have been are also common and are often infected by bacteria or protozoa. The fish shows the usual signs of discomfort, glancing off objects in the tank and sometimes dashing around wildly.

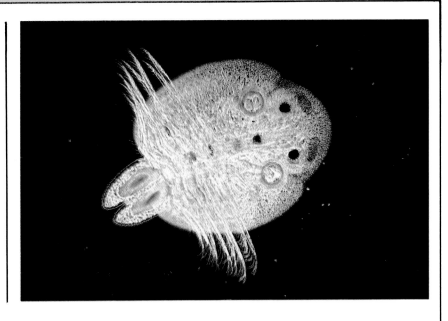

(Right) View of an *Argulus* showing the suckers and the sucking mouth.

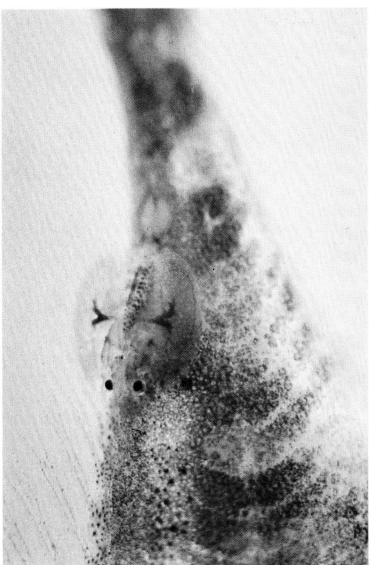

Fish lice are motile parasites, moving from place to place on the body of the fish to feed and even leaving the host to lay eggs.

The same insecticides that are recommended for *Benedenia* can be used for copepods, but rather complicated methods of treatment are deemed necessary so that the free-swimming stages can be left to die. Dulin recommends the use of DTHP in a series of two-day treatments at 1 ppm in an isolation tank repeated once a week for up to four weeks—two days in the insecticide, five days in new seawater, two days in the insecticide and so on. This is because most life cycles take less than a month and nothing survives to reinfect the aquarium to which the fishes are then returned. I would be inclined to try the same insecticidal treatment as for flukes, leaving the lower permanent dose to take care of the larval stages.

Acanthocephalan

Chaetodon octofasciatum

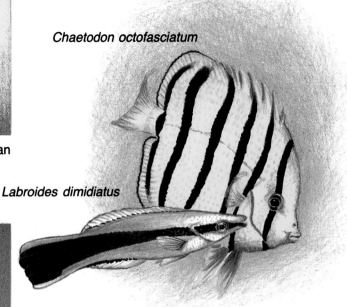

Labroides dimidiatus

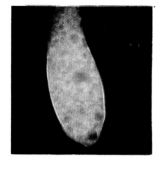

Marine Flukes

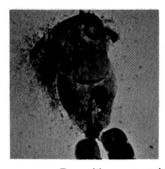

Parasitic copepod

Parasitic copepod
(*Ergasilus*)

Exophthalmus or Pop-eye
Pop-eye can be the result of disease, toxic conditions in the tank, gas bubbles, or copper poisoning. When caused by a disease such as *Ichthyosporidium* or bacterial infection, it may clear up when the disease itself is treated, otherwise you have to look for a possible non-infectious cause. One or both eyes may be affected in one or more fish, but it is unusual for more than one or two fish to show the condition at the same time. If no disease is suspected, consider the recent history of the aquarium—has it been treated with copper, has there been a recent rise in temperature that could cause gases to leave the bloodstream, or has it been over-aerated? Do any of the fishes not showing pop-eye give signs of loss of equilibrium or any other behavior that may be

due to gas bubbles in the nervous system or bodily organs? If the pop-eye is due to gas bubbles, they are likely to be present elsewhere and in other fishes and may be cutting off the blood supply in fine vessels, affecting parts of the nervous system, heart, kidneys, or other organs. Treatment depends on the probable cause. In the presence of any recognized disease, treat for the disease if it is likely to be the cause of the exophthalmus, otherwise treat for the probable cause itself. Copper should be removed promptly with activated carbon or a suitable resin. Lower the temperature as far as reasonable to aid gas absorption—usually nitrogen— and cut down on any excessive aeration, particularly that associated with any kind of high pressure filter or cartridge setup,

Cleaner wrasses will remove external parasites but cannot touch internal ones.

and hope that the fish or fishes affected may recover without further treatment. If large gas bubbles form in the eye it may take a long time for them to dissolve, as the circulation there is meager. In extreme cases, a very fine hypodermic needle may be used to suck the gas out, but skill is needed to do this successfully.
Telescope-eyed goldfish suffer from (or enjoy) genetic pop-eye, and a similar condition can be caused in goldfish by a pituitary hormone associated with the thyroid-stimulating hormone, just as in ourselves. However, a disturbance of this type has not been reported in marine fishes as far as I know.

General Malaise

A good aquarist looks critically at his tanks every day if possible. He notes anything unusual, from a slight odor to abnormal behavior of any of the fishes. In a decorative tank, are all the fishes visible, all showing normal color and movement? Do they all eat well? Is the water perfectly clear and is everything apparently normal—anemones well displayed, any live coral fully out, algae flourishing, tubeworms popping in and out? Naturally, some fishes such as wrasses may be buried temporarily, an odd anemone may be closed but looking otherwise normal, and not every shrimp or mollusk may be visible, but they should appear within a day or so. It is highly important to *look for* things going wrong and very easy to miss something nevertheless. Therefore, if anything whatever seems wrong with an aquarium, think carefully about what may be the cause if no disease is apparent.

First, check that all the equipment is functioning properly. Check the temperature, pH, nitrite level, and specific gravity. If copper has been used at any time, take a copper reading since the metal can be released from coral or gravel by abnormal chemical changes in the water or a low pH. Make sure that the undergravel filter is operating fully—no gurgling in the airlifts that would indicate poor circulation, no dead spots. Any faults found as a result of this inspection must be corrected immediately. What to do is fairly obvious in all cases except perhaps a high nitrite (or ammonia) reading. If this is found, stop feeding for a couple of days, change up to half of the water and use an ammonia-absorbing agent. Then consider what may have caused the problem—overfeeding, overcrowding, poor filtration, kitty's litter pan near the tank (ammonia), cleaning agents used in the room (could be ammonia again), and recent use of an antibiotic or other cure that may have interfered with the biological filter are all possibilities.

Second, if nothing can be found wrong with the functioning of the aquarium, look very carefully once more at the fishes. Inspect them from various angles to make quite sure that you have not missed disease, particularly velvet. If there is still no sign of disease, consider an introduced toxin. If you have copper water pipes, even if you have never purposely put copper into the tank test for copper. Copper poisoning is usually associated with new copper piping, but I myself have suffered from a breakdown of *old* copper piping that started releasing dangerous levels of the metal after being safe for years previously. This can happen because of high chlorine levels in the water, electrolytic action at joints, and several other possibilities that quite horrified me when I read up on the subject. Then consider what may have been placed in the aquarium recently—a new decoration or internal filter, new tubing, a new

This *Kyphosus* is obviously ill, but it is probable that its condition will never be accurately diagnosed. Many marine fishes suffer from passing symptoms due to stress and poor environment—when they get used to the situation they get better.

Quarantine tanks are always a good idea as they allow you to treat a sick fish away from the rest of the aquarium in subdued surroundings.

synthetic mix, and sea water collected from a new area are all possibilities. What about things going on around the household? Has there been any painting done? Has anyone used any kind of spray near the tank or near the air intake? If you smoke, do you ever have tobacco under your fingernails? Is there any possibility of someone, children for example, having put anything into the aquarium, and, if so, what could it possibly have been? A tightly covered tank guards against this to a considerable extent, but things can still happen, especially at parties.

Whether you find a specific cause for the trouble and can remove it or are still puzzled for an explanation, a partial water change and a good clean-up with renewal of any filter material other than the biological filter should help, together with light feeding for a few days and continued alertness for any further indications of trouble. Consider also whether you have

been following the recommendations about feeding, not only about quantities but about quality and variety of food. It is rather easy to become complacent and to stick to a rigid feeding regime or, worse still, to rely too much on flakes or other prepared foods. Although newcomers to the hobby are inclined to overfeed rather than the reverse, I have seen examples of semi-starvation in both freshwater and marine tanks. Everything tests out well, no ammonia, no nitrite, pH beautiful, and so on, because there isn't enough excreta to cause trouble! Only the thin and faded fishes give a clue. If lucky they may be free of disease, but it won't continue that way. If the aquarist is not used to seeing healthy, well-fed fishes, the gradual deterioration may not be noticed until it gets severe.

The symptoms of various non-infectious conditions are very diverse, but they are discussed below as a guide to what may be wrong.

Poisons generated within the tank, especially ammonia, cause rapid respiration, atypical swimming movements, attempts to jump out of the water, color changes either darker or lighter than normal, and eventually various infections and death. Ammonia, hydrogen sulphide, and phenols are the most common internally produced toxins, in about that order, with ammonia by far the most frequent culprit. Hydrogen sulphide (H_2S) is a gas smelling of rotten eggs generated when anaerobic bacteria get to work in unaerated pockets of gravel or in a really dirty filter that is working badly or has been turned off for some period. It is deadly at less than 1 ppm, combining with the hemoglobin in the blood and preventing it from carrying oxygen to the tissues. Ammonia has no odor in the concentrations we are considering, but hydrogen sulphide does, as indicated. Phenols may be formed from uneaten food, especially tubificid worms and decaying algae. They may also be released from cheap plastics. Phenols are also highly toxic, but their existence in the tank can only be suspected from the history of feeding, conditions in the tank, and the visible state of its contents.

Introduced poisons may cause very similar symptoms, but the check you will have made of ammonia and nitrite levels, pH, etc., and the smell of the tank will enable the exclusion of these factors plus hydrogen sulphide as a cause. A commonly introduced poison is a detergent on unwashed hands or hands washed with a detergent and not properly rinsed, on dishes used for conveying food to the fishes, or on anything placed in the tank. Detergents are doubly dangerous—poisonous in

themselves and also by stripping
the fishes of their mucous coats,
impairing skin and gill resistance
to infection or to other toxins.
The benzosulphonates commonly
used are toxic in a few ppm and
dangerous if present for long
periods at probably much lower
concentrations. Don't be misled
by advertisements showing dishes
drying without being rinsed clean
of a detergent; they are probably
bad for you as well as being
potentially lethal to a fish. Soap is
also bad but is less likely to be a
cause of trouble because people
seem to be more aware of the
danger. Metallic poisons may be
introduced from unprotected
furnishings, the lighting hood
being the most frequent. Not
only may there be drip-back from

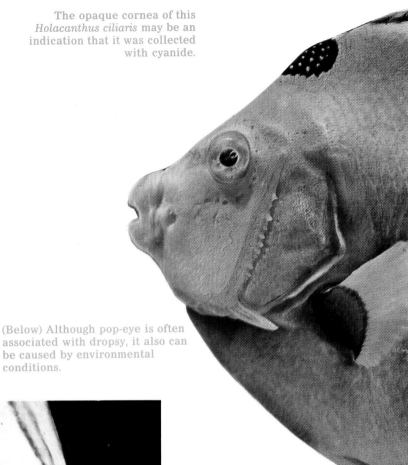

The opaque cornea of this
Holacanthus ciliaris may be an
indication that it was collected
with cyanide.

(Below) Although pop-eye is often
associated with dropsy, it also can
be caused by environmental
conditions.

a hood, but salt deposits have a
habit of creeping everywhere and
can attack the flanges where the
hood rests on the edges of the
tank, thence to fall back onto the
glass covers and perhaps be
washed back into the tank. A
degree of contamination from
iron (as in stainless steel) is
tolerable, but much more than is
naturally present in sea water
becomes toxic. The really
dangerous metals are mercury,
silver, zinc, cadmium, copper,
and lead—hence no mercury
thermometers to get broken, no
galvanized iron as a tank frame
(luckily now obsolete), no lead
sinkers, and absolutely no
metallic ornaments that may be
safe in fresh water.

131

A small tumor on the nape of a gourami, *Colisa*.

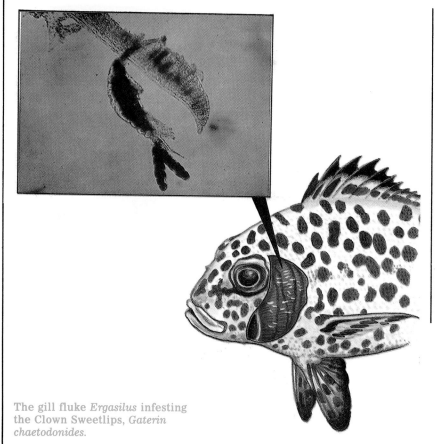

The gill fluke *Ergasilus* infesting the Clown Sweetlips, *Gaterin chaetodonides*.

Tumors

Tumors and spinal abnormalities are often genetic in origin and thus of no danger to other fishes. If no other disease is recognized and a fish with a growth doesn't appear distressed, it may be safe to leave it in the aquarium, but be sure that the problem is not caused by *Ichthyosporidium* or some other infection. Since this may be difficult and because an affected fish doesn't usually look very nice, it is best to get rid of it. Although rare in marine fishes in the aquarium, tumors of the thyroid gland have been noticed in older specimens kept for some time in water low in iodine.

All early aquaria, whether freshwater or marine, were of necessity natural system tanks without aeration or filtration of any kind. Even in the days of Innes (1917 onwards) it was normal to keep freshwater fishes in unfiltered, unaerated tanks, although methods for employing both accessories were being developed. Today, nobody sells you an aquarium without offering a filter and aeration, combined or independent of one another, although a freshwater aquarium keeper can do without either quite successfully and keep about as many fishes as he usually does. It is different with marine fishes, but not all that different, as we shall see.

Cross-section of a typical reef showing the position of the inhabitants. Clockwise from left: Moray eels, lionfishes, squirrelfishes, chaetodons, anemonefishes, mandarins, clingfishes, bubble coral, carpet anemones, damsels, hard corals.

The Balanced Aquarium

As early as 1819, a chemist called Brande pointed out that "Fishes breathe the air which is dissolved in water, they therefore soon deprive it of its oxygen, the place of which is supplied by carbonic acid; this is in many instances decomposed by aquatic vegetables, which restore oxygen." But it wasn't until 1850 that Warrington, another chemist, reported that with plants growing in it, the aquarium did not need water changes. Gosse, at around the same time, made similar observations and extended them to include marine as well as freshwater aquaria.

However, the idea of the "balanced aquarium" in which plants renew the oxygen needed by the animal life was gradually abandoned. In the words of Atz in 1949, nearly 100 years later, "As soon as the slightest deficiency in oxygen exists in a tank, oxygen from the atmosphere passes into solution to make it up. Similarly, if an excess is produced by plants, under the influence of bright light, this quickly passes off into the air." He went on to point out that carbon dioxide moves much more sluggishly and that an excess of it in an aquarium takes an appreciable time to pass off.

Atz was of course correct in his description of the behavior of the two gases, but he did not adequately discuss the other functions of plants in absorbing vast amounts of organic material from the water as well as utilizing carbon dioxide when under adequate illumination. They may not contribute a great deal to oxygen concentration in the average tank, but they do remove a lot of unwanted materials, and that is why they help to keep the water sweet. "What" you may ask "has all this to do with the natural system?" The answer is a very great deal, since the system employs growing algae, not only in the free state, but also as components of the many corals and anemones that act as their

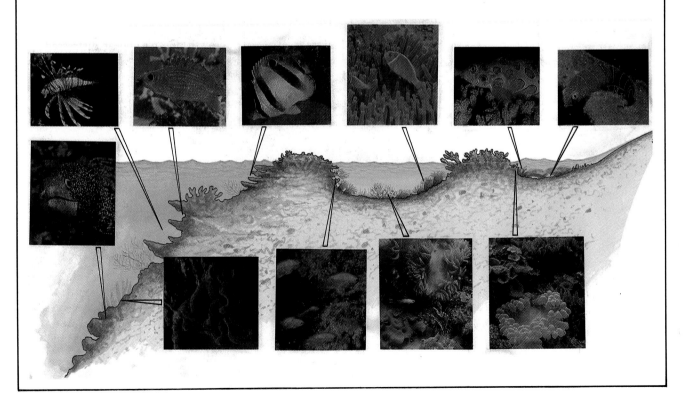

hosts.

An aquarium must have the breakdown products of its living contents dealt with in one way or another. This may be achieved by frequent water changes, by physical or chemical filters, by sufficient surfaces on which nitrifying bacteria can grow, by the presence of healthy plants, or by combinations of these. The old-fashioned fish bowl depends only on water changes; many marine aquaria depend on the first two or three and some on all four. An unfiltered aquarium doesn't get too much help from nitrifying bacteria and it isn't going to work as a "natural" system unless it gets help from algae.

Functions of Algae

In adequate light, growing algae, single- or multicelled, produce carbohydrates from carbon dioxide and water and from these they can build fats, but not proteins. Proteins contain nitrogen, available in the aquarium from waste products, primarily ammonia, nitrites, and nitrates. Some algae can utilize all three of these, but all can manage at least nitrates. Thus the algae, with the aid of various trace elements, can be very important cleaners of the water, and if a lavish growth is encouraged and some is periodically removed, the net result is removal of waste products from the tank. At worst, the waste products are reconverted to healthy plant material and do not contaminate the water. If eaten by fishes or

The natural appearance of this scene is due not so much to the fishes as to the algae and invertebrates in the background.

other animals, they will be released but again recirculated, reducing the overall rate of contamination. Otherwise comparable tanks with and without growing algae show an approximate halving of the nitrate level, for example.

Algae living in the tissues of corals and anemones do exactly the same *within* their hosts, if adequately lit. They absorb the waste products of the host in manufacturing their own substance and eventually feed the host as well. Otherwise unfed anemones carrying a normal population of zoochlorellae and zooxanthellae (symbiotic algae) have been seen to grow in size in the aquarium. This implies that their symbiotic algae are utilizing more than the anemones' own waste products and must be assumed to be using some from the surrounding water as well. Insofar as this is true, such anemones are, with the aid of their algae, helping in purifying the aquarium water, and so presumably are some of the corals— *in adequate light*. In fact, it has been shown that in the presence of a coral's digestive enzymes, the algae within it are

"leaky" and give up to 80% of their production to the coral, mainly in the form of glycerol and acetate. One feature of a natural system aquarium is the presence of growing algae on rocks, etc., or in animal tissues, which must be well illuminated. We can begin to see why the natural system works.

Various one-celled green algae (zoochlorellae) and yellow or brown algae (zooxanthellae) are associated (symbiotic) with protozoa, sponges, coelenterates

The establishment of a variety of invertebrates in the marine aquarium can be a time-consuming task. Starfishes, by the way, are usually considered unsafe in a small aquarium.

and other phyla. They are prominent in reef corals and some anemones and are responsible for some of their color, particularly the zoochlorellae. If they die, the animal may no longer flourish, even if fed, since its own waste products are no longer disposed of by the algae and may not escape quickly enough into the water.

Pigmented anemones and other cnidarians such as soft corals and hard corals depend on algae contained in the cells for at least a part of their food. If the algae get insufficient light, the host will suffer or die.

Starting the System

There are several ways to start off a natural system tank. The essential common feature is that there is no chemical or other filtration except possibly biological filtration with an undergravel filter, but this is usually omitted and there is just a thin layer of sand or gravel. Aeration may be very crude, by a bare air-line pinned down under a rock or coral or by conventional airstones. The modern natural system for marine aquaria was introduced by Lee Chin Eng and in his hands underwent a degree of development. Others trying to follow his lead had very diverse experiences, mostly reporting various degrees of failure. This was largely because Lee did not publish much about his methods, so it was impossible to do exactly as he did, and that was the result of a good deal of experimentation on his part. It was also impossible for many aquarists to follow Lee's example, even if they knew what to do, because Lee took his water and his aquarium contents straight from the ocean to his tanks with only a few hours' interval at most. This makes a

(Above) Before more delicate invertebrates such as shrimp are added, the water chemistry must be stabilzed. If the ammonia and other wastes are not fully transformed, many invertebrates will die. Although anemonefishes are a type of damselfish (below), they also are quite sensitive to poor water chemistry.

very big difference in the condition of "living" rock (rock with algae, tubeworms, anemones, and other invertebrates in or on it) and some corals, the core of the method.

Method 1 If you are fortunate enough to have ready access to the seashore and can transport sea water in sufficient quantity, follow Lee's method even if your local sea is temperate rather than tropical. Really cold shores will have a sufficiently different flora and fauna to make adaptation to a tropical tank unlikely, but you can have a spectacular cold-water natural tank as some of the anemones and other invertebrates are quite as colorful as tropicals, but the anemones will not be carrying zooxanthellae. Fill the tank with sufficient sea water, remembering that it will have to house quite a volume of contents; get it going nicely with an airstone or two, depending on its size; and heat it to around 78°F (or leave it cold if you are to have a cold-water setup). Don't subject the sea water to dark storage, use it new. Any

infectious organisms to fishes will not matter at this stage. When everything is in order, put a layer of preferably coral sand not more than ½" thick on the bottom, and start the ball rolling. Aim for a nice mix of living rock, which you may have to chisel off from shore pools (if legal), and coral if available. Put as much living rock or coral as you can safely transport without suffocating it—damp packing in plastic bags with some air is better than immersion in water even if you can oxygenate.

can be achieved quite rapidly or by degrees as opportunity presents. If, despite all care, there is some decay or die-off in transport, put up with the stink, if there is one, as it will disappear, but turn up aeration to the full temporarily. There can be a mixture of live and properly cured dead coral if you wish or are forced to it; it will all blend in eventually. Live corals to be added are preferably those like *Fungia*, consisting of a single large polyp, or whole coral heads collected without breaking them

off from other living material. Soft corals or those from deep water without indwelling algae can be placed in the darkest spots, as they do not require bright light to flourish.

At this point, wait for developments. Not very much in the way of animal life may be seen at first, although the algae should flourish as long as enough light is provided, at least 30 watts of suitable fluorescent tubes per foot run of tank if the tank is more than 15" deep, at least 20 watts if it is shallower. Do not attempt the natural system in very small tanks unless you wish to present a series of virtually individual specimens. After a week or more, particularly if the living rock came from semi-tropical or tropical areas, things should begin to happen. Crustaceans, worms, sea-stars and other echinoderms, nudibranchs or sea hares, and medusae may all manifest themselves, together with hatches of plankton that may temporarily cloud the water and help feed the other inhabitants. How much of this occurs is largely fortuitous but is dependent on the living rock, not much on the coral which may, however, have some passengers with it. Later you can add anything you want that's missing from the initial mix, but give it time to develop on its own for a month or two. Get the algae, especially the higher algae, growing and add some if necessary. Feed any large anemones lightly if they seem to need it, but don't unnecessarily pollute the aquarium. Lots of tiny critters and some of the larger ones that "come out of the woodwork"—in this case the living rock—will perish, but plenty should survive.

Eventually cover the back and sides quite fully with rock and coral, or fairly fully if rock alone is available at that stage. Hide the air-lines and stones behind this, about one airstone per 2 square feet of surface area being adequate. If you have to purchase coral, this can be added as you like as this stage of development

Lee Chin Eng was the "father" of the modern natural system of keeping marine fishes. Many of his ideas have been incorporated into the techniques and equipment used in today's mini-reefs.

Method 2 A *very good* brand of synthetic mix may be used instead of natural sea water, but it must have guaranteed trace elements essential to algal culture and to invertebrates. Choose a brand that provides a full analysis. Very pure tap water should also be available; if it isn't, try hard to get guaranteed demineralized or glass-distilled water. Ordinary distilled water may be distilled over copper and is then loaded with it, so be very careful! If in desperation, use tap water as long as you drink it yourself and filter it thoroughly with carbon or resins for a few days before introducing anything else to the tank, and do the same with all replacement water in the future. If there is any question of chloramine, treat as necessary before adding the salt mix. Resins of one kind or another may be slow to remove some contaminants, especially heavy metals in very low concentrations, but high enough to be dangerous.

Synthetic sea water is devoid of the plankton present in ocean water, although it will not be sterile. Instead, like stored sea water, it will be loaded to a greater or lesser extent with bacteria. There will be no eventual difference once the living rock, etc. have been introduced, but events may proceed a little slower. Nothing will come from the water itself, as items from the plankton that might otherwise develop into little plants or animals aren't there. Once the aquarium has been equilibrated the contents may be introduced and off we go as before.

Method 3 This is a way of starting up if you have to purchase rocks and corals from a dealer, in whose tanks they will have lost much of what they might have had. How much remains depends on the condition of the living rock when he received it. If it was packed wet, it pays to ask him to let you be present when a batch is received and to take your pick before anything else is done with it. You will not then lose a great deal to the dealer's tanks that might otherwise have gone into yours, and if you ask him to let you have the drainings in the shipment bags as well, you may collect a lot of very interesting critters.

Often, because it travels cheaper that way, the rock is sent almost dry and by the time it arrives is half dead or worse. The dealer will then place it in empty undergravel filtered tanks until the smell subsides and perhaps the nitrite readings with it, although the latter measure the efficiency of the undergravel filter rather than the state of the rock. After a week or two each piece is shaken thoroughly under water and placed in a tank for sale.

It is a wonder that, after all

For many years a cheap damsel-fish has been the test animal of choice to determine if the water in a new marine aquarium will support life. If the damsel lives, the water is probably safe.

that, the living rock isn't completely dead, but it won't be, if it was originally worth having. Some "living rock" on offer never was much alive. From a good rock things will develop, but only in time and in the right conditions. There will be a few surviving creatures in cracks and holes, spores will hatch out, and remnants of algae will grow again. Bacteria probably will be abundant already and will start up nitrifying activity in the tank. Marine plants can be added at any stage to hasten things up, as these are available commercially. Progress will naturally be slower than by the first two methods because much less life is there to start with, but the tank will catch up, with in all probability a more limited range of inhabitants until you start adding them, and perhaps even then.

Method 4 This method cheats, but is beginning to show its merits. An undergravel filter is set up as usual, but the tank is thereafter treated as a natural system tank. The rate of flow is kept minimal, enough to keep the biological filter sweet but not as brisk as would normally be practiced, so as to keep the plankton up in the water as much as possible. Some aquarists also use less substrate than normal, say only 1"—2". The rate of flow of water down into the filter bed is quite small, as a simple calculation will show. Suppose we are turning the water over in a tank 18" deep, at the rate of two tank volumes per hour. This means that 36" of water slips down over the whole area of the filter per hour, or 0.6" per minute. Zooplankton can cope with this, but some of the phytoplankton may get trapped. Does it matter?

Actually, with a fairly thin filter bed, some aquarists turn the filter off semi-permanently and leave it there as an emergency measure should there be a threatened fouling or wipe-out. This seems very hazardous, but they get away with it! It seems much the best to keep a gentle flow going and increase the nitrifying capacity of the

Living rock is actually a varied assortment of sessile invertebrates and their growth stages. Clockwise from top right: zoanthids, hard corals, hydroids, tubeworms, corals, sponges.

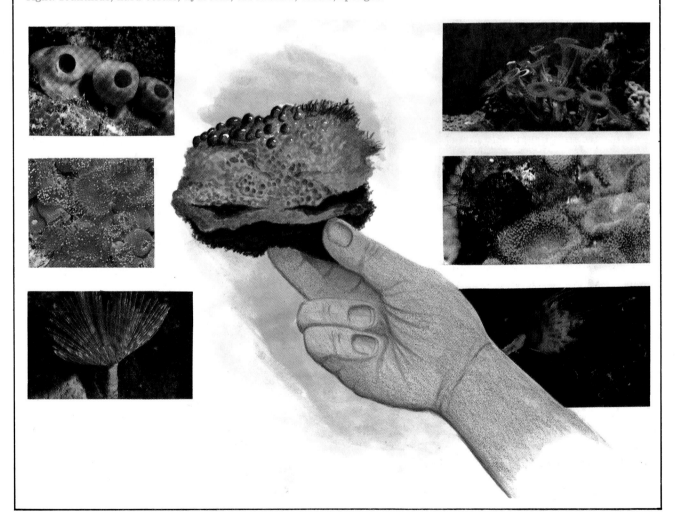

The successful mini-reef is a blaze of color, much of it derived from the plants and invertebrates.

aquarium considerably without any danger either to the system or the biological filter. Lee agreed that such a tank works very well, but I had the impression that he felt that it was cheating (unfortunately he is now dead). Of course, we have only to add a carbon filter or any other devices and we are back to normal, but then we shall lose the plankton.

Later Developments (All Methods)
After a satisfactory start, living invertebrates may be introduced at the rate of about one piece or specimen per week. The more living rock or algae-containing anemones or corals present, the greater the eventual capacity of the tank. These are the equivalent of a biological filter and must support not only themselves but the other animal life present. Nobody has presented other than "guesstimates" of the capacity of such tanks, but, based on experience and observation, they

can be pretty densely populated. The more living rock present, the more plankton will be produced from its inhabitants, and swarms of it will occur. In the absence of much living rock, with corals instead, much less of this happens and there is a tendency for the tank to be able to support fewer animals—be they invertebrates or fishes. The need for feeding will also vary with the amount of living rock, the more of it present, the less the need for external feeding as food in the form mainly of plankton and algae is being produced. Corals and anemones do not feed the tank unless creatures that eat them are present. Reproduction in these coelenterates is not going to be frequent enough to do much; corals will probably never contribute at all.
Fishes can be introduced after everything else is deemed to be going well and the tank appears ready to receive them. Lee put up to a dozen ¾" to 1¼" fishes into a 36" tank—quite a number.

We must ponder, however, on the biomass of fishes and various invertebrates and on how much they consume. This determines their eventual production of nitrogenous compounds that have to be dealt with in the tank. A fish is about 20 to 30% dry weight—i.e., 70% to 80% of it is water. Some invertebrates like crustaceans may approach this dry weight percentage, even omitting the shell, but many invertebrates have much less living tissue or have much more water in their composition or both. Think of anemones, corals, jellyfish, and other medusae, all of which are nearly all water, excluding any calcareous or other skeleton. In terms of normal visible bulk, invertebrates will on the whole contribute much less than fishes to the waste matter of the system and a tank full of them may have a lower biomass than a few fishes. If a tank could normally hold, say, 20 small fishes, it could well, and in fact *can,* hold about 12 small fishes and a mass of invertebrates, even if we ignore the capacity of some invertebrates to support themselves. If you wish the tank itself to be a main source of food for the fishes, fewer than Lee kept must suffice—perhaps only three. But there is no need to be so frugal, so why not enjoy the fishes as well as the invertebrates?

Feeding

Lee fed his fishes and invertebrates quite heavily compared with the practices of others. He bred small animalcules and crustaceans in a special vat, collected plankton from the ocean, and also fed mosquito larvae and anything else suitable, but rarely gave dry or frozen foods. This in part reflected his relatively heavy loads of fishes and small amounts of living rock, so that his in-tank production of food was rather low. There would be no objection to feeding the fishes in a natural system just as heavily as in any other, but if in-tank production is high relative to the number of fishes, the need to do this is reduced. Thus, with only a few fishes per tank you can get away with occasional feeds, the frequency depending on just what is happening in the tank. One noted keeper feeds his fishes only weekly, with a few mysid shrimp each or the equivalent of other foods, and also puts shelled brine shrimp eggs into the tank "occasionally". If his fishes flourish, there must be a very good plankton production rate in his aquaria. My own experience has been between the two extremes, and I don't think that I have ever had a tank that produced sufficient food, in my own estimation, even for a few fishes on a regular basis— sometimes, yes, but not often enough. In part, it all depends on what you expect of your fishes— I like to see a healthy growth rate, yet a half-starved fish on an otherwise well-balanced diet can stay healthy but not grow much, if that's OK with the owner. Fishes can even breed at a fraction of their normal mature size if kept undernourished; they are very accommodating pets.

Since anemonefishes feed during the day and sleep in the anemone at night, it is often difficult to feed their anemones.

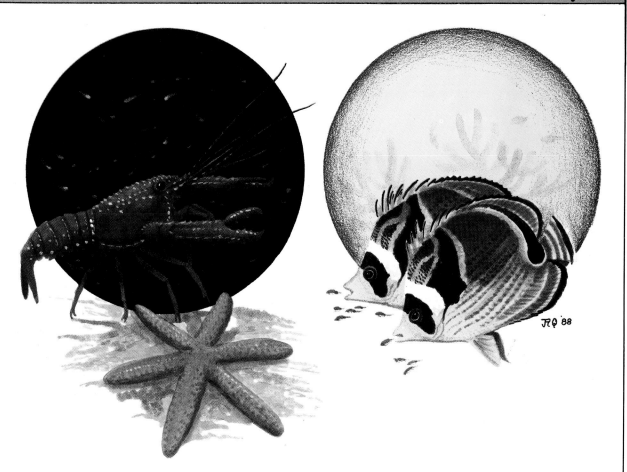

When feeding your pets, it is important to remember that some are diurnal (day-active) and others nocturnal (night-active). Many crustaceans and invertebrates feed only at night or at least in reduced or red light, while many of the more brightly colored reef fishes are active feeders during the day. Predatory crustaceans commonly are active at night, which also means they can attack sleeping diurnal fishes.

Not only the fishes but also the invertebrates need feeding. Just how and with what depend on which are present. It is good to have a lot of filter-feeders in the tank, and these need plankton, which in part they will get from the tank, but again probably not sufficient. They can be fed on newly hatched brine shrimp, liquid fry foods, and fine dry fry foods, any of the standard early foods for egg-laying fish fry, in fact. Any plankton you can collect or obtain will also be appreciated. The more, within reason, you feed them, the more at least some of them will produce for their companions and probably themselves to eat. But be cautious—an unfiltered natural system tank can foul up easily if overcharged with artificial foods.

Feeding crabs and shrimp is no problem as long as you make sure

that they get some food. Nighttime feeding is a solution, as the fishes are then quiescent while the crustaceans tend to be active. Anything usually fed to the fishes will do fine. Large crabs, hermit or otherwise, are messy eaters and also turn over the substrate and even quite large rocks, so be cautious about introducing them. They are good scavengers, though, and can be very useful in a large tank if kept within bounds. Young lobsters seem quite intelligent and quickly learn to fish down into an anemone and remove whatever it has ingested, a trick I have never seen a crab learn to do.

Various mollusks can largely be left to their own devices. The filter-feeders will get their share when that class of creature is fed, others will crop the algae from glass and rocks, and yet others are carnivorous and best avoided

143

as they may attack other mollusks. Cuttlefish and octopuses are also unwelcome in a community tank, unless they are very tiny, as they attack many different animals.

Small anemones and corals are best fed with a baster so that you can squirt the food gently over their tentacles. Feed any reasonably small material to them, such as newly hatched or immature brine shrimp (alive or dead), finely chopped animal material of almost any kind, even fine dry food after soaking in water first. Only feed them when they are fully expanded.

Large anemones, even those with flourishing zoochorellae or zooxanthellae, need feeding fairly often. About twice a week is best with smaller pieces of food than

Perhaps the best way to feed most anemones is to actually put a small piece of food into the mouth using planting tongs or a similar device. This way the sessile animals are not dependent on accidentally having food pass over them.

they are actually capable of ingesting, so as not to overtax the animal or the aquarium. Those without anemonefishes living with them can be fed at night with the aid of dim lights, but the fish or fishes associated with others present a problem, both by night and day! At night the fish is tucked into the anemone. It can be disturbed, but I don't like doing it. In the daytime it is more likely to steal the food than to give it to the anemone as so

many writers describe. I train my large anemones to accept pieces of shrimp, fish, etc., on the end of a probe and to allow me to thrust the food into their mouths. At first it may be necessary to brush it against the tentacles a few times to stimulate a feeding response, but even lowly anemones quickly learn to accept it straight off—to the confusion of the anemonefishes that cannot then steal it. Some species of anemonefishes help to feed their anemones more than others do, and even then probably more by accident than design, as they appear to use the anemone as a hiding place for food as well as a refuge for themselves and if it is lucky the anemone itself gets some.

Starfishes vary in their requirements. Some will open live mussels or other bivalve mollusks and insert their stomachs to digest the prey, some will deal with pieces of animal material, and others feed on algae in the tank. Brittlestars and featherstars are often quite active feeders, feeling around at feeding time for pieces of nourishment, curling their arms around them and dragging them into the scenery where they themselves usually lurk.

Maintenance

The less a natural system aquarium is disturbed, the better. There is no cleaning of decorations, rocks and so forth, no regular removal of mulm; only the front glass is kept clear for viewing. Water changes are needed, but with less frequency than in an ordinary setup. As little as 10% every two months may be sufficient. It is advisable to add a trace element supplement as recommended by the manufacturer or perhaps a somewhat higher dose, as this type of system extracts more from

the water than is usual; both
growing algae and invertebrates
are doing it.

Algae may need culling from
time to time and mulm may need
removal every few months. Some
aquarists use a protein skimmer
for a few hours occasionally; such
a procedure will help clean the
water and will not remove much
plankton. If the water colors up
appreciably, a skimmer is about
the only method for removing as
much of the color as possible that
can be used, as any kind of
filtration is out. As a last-ditch
measure, it is possible to make a
few rapid 30 or 40% water
changes, say one every two days
for a week. The aquarium will
soon renew any losses of
plankton. Three 30% water
changes will renew two-thirds of
the water; three 40% changes will
renew four-fifths. It is inadvisable
to shock the inhabitants with
more than a 40% change, even
with natural sea water.

HIGHER ALGAE

Three groups of algae make up
the vast majority of seaweeds,
which are multicellular or
multinucleate algae with a
relatively large plant body called
the thallus. The groups are
named from the predominant
pigments they contain. The green
algae have only chlorophyll and
so nearly always look green. The
red and brown algae have

chlorophyll plus various other pigments that may mask the chlorophyll and give a red, brown, purple, etc., look. However, particularly the "red" algae may look almost any color and are hard to classify from their superficial appearance. Most of the aquarium seaweeds so far popularized are green algae; reds and browns appear on living rock but are usually difficult to identify.

Green algae (Chlorophyta): With few exceptions, algae produce spores that grow up to form a new plant. In the green algae, these spores are usually motile, flagellated zoospores that settle down away from the parent plant. The new plant, called a gametophyte, produces gametes (reproductive cells) and is thus the sexual generation. In most marine green algae the gametes are flagellated, like spermatozoa, and may be equal or unequal in size. Like the gametophyte itself, these gametes possess only one set of chromosomes and are thus said to be *haploid*. When they meet and fuse a *zygote* with a double set of chromosomes is

formed and is thus *diploid*. This zygote grows up into a new plant that is a representative of the asexual sporophyte generation and produces spores that repeat the whole process. Sometimes, as in *Ulva*, the sea lettuce, both generations look alike; in other cases they do not. The usual seaweed described, when there is a difference between the appearance of the generations, is the sporophyte generation. This may also reproduce by fragmentation, especially in some of the simpler filamentous forms like *Caulerpa*, but spore formation is the general rule. The various *Caulerpa* species are popular because of their readiness to fragment and continue to grow in the aquarium, in contrast to most other algae.

Brown algae (Phaeophyta): These algae often have very complicated life cycles. Most of them have alternation of generations as just described for the green algae. The familiar plant is the sporophyte, and in the Fucales, to which many common rock weeds belong, the gametophyte generation has

disappeared as a separate entity. Instead, there are eggs contained in conceptacles on the thallus, fertilized by sperm also produced from the thallus—a development similar to that common to many advanced plants and animals. In those brown algae that produce gametophytes they are usually tiny in comparison with the frequently very large sporophytes, but there are orders (Ectocarpales and Dictyotales) in which both generations look alike.

Red algae (Rhodophyta): These algae were no doubt created especially to confuse the student, as they are not only often not red, but have in the majority of cases three generations, not two. There are sporophyte and gametophyte generations that usually look alike and a carposporophyte generation that remains attached to the gametophyte generation. The sporophyte produces non-motile spores that produce separate male and female plants, and these in turn produce non-motile sex cells. The male cells are set free, stick to a female plant's sex organ (rather like pollen), and fertilize the "egg" cells that then grow into the carposporophyte generation in or on the female plant. This third generation produces spores, the carpospores, that develop into sporophytes. Finally, as an exception to the usual rule, some red algae have large gametophytes and tiny sporophytes.

These complications in the life history of algae have only been discovered by painstaking study in the laboratory after culturing the plants artificially and demonstrate that these marine algae are by no means primitive forms and have evolved in a most complex manner. The situation is mentioned briefly not only out of

Caulerpa racemosa.

interest, but to indicate what you might expect of algae in the aquarium. Some will reproduce easily by fragmentation, but the majority must go through the various stages of their reproductive cycles in order to provide new plants. A single sporophyte may reproduce, a single gametophyte may not, but even if the sporophyte does its act, it will produce gemetophytes that will most often be tiny and unrecognizable as the same species of plant. If you are lucky, another sporophyte generation will follow and you will see the same alga again.

Types of Green Algae

Of all the green algae, the genus *Caulerpa*, of which there are many species, is the most adaptable to the aquarium and the most frequently seen. It is a member of the family Caulerpaceae, which contains most of the other genera of popular aquarium algae, namely *Halimeda, Udotea, Penicillus,* and *Rhipocephalus*. *Caulerpa prolifera* has flat green fronds with a root-like branching holdfast by which it attaches to rocks, corals or the substrate. It is very easy to grow in adequate light. It is found from Brazil to North Carolina and is also produced commercially. Almost any fragment will reproduce and give rise to the runner-like rhizome from which the holdfasts and fronds arise.

Caulerpa asmeadii, C. mexicana, C. lanuginosa, C. crassifolia and *C. sertularioides* are all more attractive, fern-like species in which the fronds are dissected into opposite pinnate branches (like a feather). All occur on tropical Atlantic coasts. *Caulerpa paspaloides* is even more ornate, as its fronds are further dissected into secondary

A fern-like *Caulerpa*, possibly *C. ashmeadii.*

(Above) *Stenopus hispidus* and *Caulerpa.* (Below) The anemone *Bartholomea annulata.*

Caulerpa cf. mexicana.

(Above) *Caulerpa prolifera.* (Below) *C. racemosa.*

branches. It is also a tropical Atlantic species. *Caulerpa racemosa* and *C. peltata* have rounded knobs on branching fronds and sometimes resemble a bunch of grapes. *C. verticillata* has finely divided tufts rather like a shaving brush, while *C. cupressoides* looks like a long-stemmed cactus. They are found on the coasts of Florida. Most of the *Caulerpa* species can be quite variable in appearance, depending on the particular conditions in which they are grown. The descriptions given are of those found in the wild and in optimum culture conditions.

The other members of the Caulerpaceae mentioned are all calcareous, meaning that calcium is deposited in or on some part of the thallus. They do not reproduce from fragments as do the *Caulerpa* species and should be acquired with the holdfast intact, preferably attached to something. Species from all genera are to be found on the warmer Atlantic coasts; indeed, most occur practically all over the sub-tropical and tropical world.

Halimeda species are all cactus-like, with a fibrous holdfast and calcified segments joined by flexible connections. *H. discoidea* has nearly circular plates nearly an inch across and is called "baby bows".

Udotea species have a flattened thallus composed of moderately calcified filaments and hence are known as green sea fans. *U. flabellum* and *U. cyanthiformis* are attractive species. *Penicillus* species have a terminal tuft of free filaments, giving them the names Neptune's shaving brush and merman's shaving brush. *P. capitatus* grows to about 5″ and is found in sandy or muddy areas. *Rhipocephalus* has a terminal tuft with filaments united into tiny blades, but some specimens look very like a *Penicillus*. There are only two species in the genus, *R. phoenix* and *R. oblongus*; both are confined to the West Indies.

The family Ulvaceae is represented by the sea lettuces, *Ulva* and *Monostroma*. Many species of each are found worldwide, and they differ only in having either two or one cell layers in the thallus, respectively. The hair algae *Enteromorpha* and *Blidingia* belong in this family and again are worldwide with many species. These are algae most likely to appear spontaneously in the marine aquarium. The family Dasycladaceae is also large and widely represented, having in it the attractive *Cymopolia barbata*, which looks like a branching string of beads tufted at each tip.

Types of Brown Algae

Caution is needed in acquiring brown algae that do not appear and thrive spontaneously on living rock, particularly any large plants with a fleshy thallus. This is because they are very likely to degenerate in the aquarium and cause havoc, releasing slime, toxins, and color into the water. Confine experiments to small, thin, calcified or wiry specimens that can do little harm.

In the family Ectocarpaceae is the genus *Ectocarpus*, with many hard-to-identify species. They are small hair algae often found growing on other algae and in the aquarium tending to smother them. A more attractive small alga is *Padina*, each frond of which is a little fan, usually with splits in it. *Padina vickersiae* grows to about 3″ and is lightly calcified; *P. sanctae-crucis* is strongly calcified. In the same family (Dictyotaceae) and safe in the aquarium are *Dictyota* and *Dilophus*, both genera containing many species with a thin flattened thallus looking like branched, brown *Ulva*. *Colpomenia* is a hollow alga and is

(Above and below) An encrusting green alga, possibly *Dictyosphaeria*.

Caulerpa cf. *ashmeadii*.

A brightly colored branched sponge.

(Above) *Caulerpa* cf. *mexicana*. (Below) *Stenopus hispidus*.

often found floating. All these genera are virtually worldwide, but not, of course, the individual species.

Types of Red Algae

The red seaweeds are mostly small and hardy, forming quite suitable aquarium plants less likely to foul the water than the browns. They flourish in poorer light than is needed by the green algae and inhabit lower zones than the others. Two genera of hair algae, *Bangia* and *Polysiphonia*, are exceptions to this rule, but both are brown or purplish in color rather than red and occur on rocks in shallow tide pools. *Bangia* species are found in thick matted patches an inch or 2 in length. *Polysiphonia*, of which there are over 200 species, some with branching thalli, occurs in rings with a dead center. *Pterocladia capillacea* also inhabits intertidal regions in many parts of the world and is an attractive, branching red to russet-brown plant 3 or 4 inches long. *Laurencia* is another branching, fern-like red alga with a number of attractive species, but these are found at the lowest tidal levels.

The limey, coralline species of red algae form bright pink encrustations on rocks, familiar on "living" rock, while those of a more typically plant-like form are usually small, jointed, branching, and hard to the touch. Dried in the sun, they stand out conspicuously as white skeletons. Common species belong to the genera *Corallina*, *Jania*, and *Amphiroa*, all in the family Corallinaceae. *Corallina officinalis* is an example of a 3- to 4-inch Pacific coast species that is quite robust, while *Jania* and *Amphiroa* species tend to be delicate, spindly, and mixed up with other small algae rather than occurring as individual growths.

In the family Champiaceae are found various species of *Champia*. *Champia compressa*, from Australia, is a very small red alga only up to about ½″ long, occurring in tufts that shine under water, but only under water, with a brilliant blue-green iridescence. Out of water it appears translucent red or brown. American species such as *C. paroula* look very similar to *C. compressa* in illustrations, but I do not know if they shine under water like the Australian species.

Suitable Invertebrates

Almost any invertebrate that can be fed and housed successfully can be kept in a natural system tank; the main problem is compatibility. Creatures like corals and tube worms that do not move around are no problem as long as nothing is introduced that feeds on them. Filter-feeders in general are not predatory, so clams, mussels, scallops, and sea cucumbers, even if some of them do move around, are no problem. It is when we come to the frank predators that we have to be selective and calculate the chances of a successful community. This applies to the survival of algae as well as that of animals.

Most anemones are safe, as they do not seek out prey and the great majority of animals learn to avoid them. The odd loss will occur, but it is exceptional *except* in the case of the highly poisonous *Cerianthus* species. These tube anemones are not closely related to the others and are very attractive looking, with pale pastel colors. They grow to several inches in diameter, with a long papery tube normally buried in the substrate. Fishes and shrimps can touch most anemones without serious harm, but a *Cerianthus* can kill them.

A young giant clam, *Tridacna*.

(Above) *Caulerpa* cf. *prolifera*. (Below) A sabellid tubeworm.

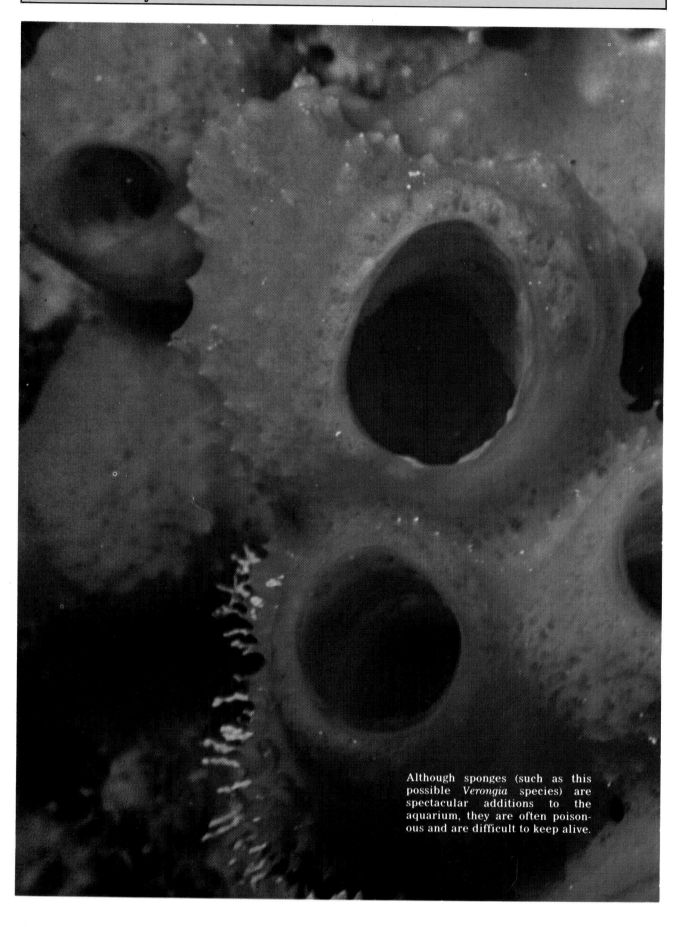

Although sponges (such as this possible *Verongia* species) are spectacular additions to the aquarium, they are often poisonous and are difficult to keep alive.

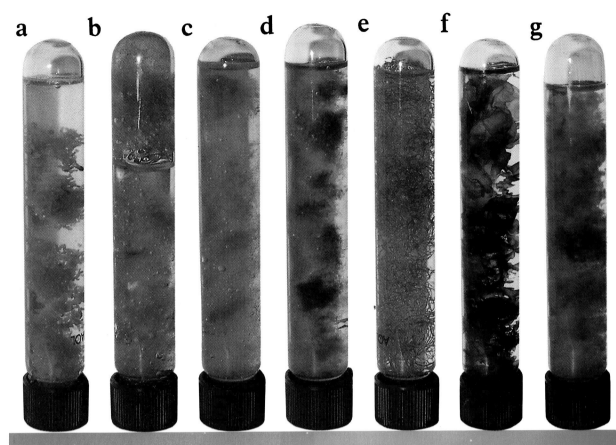

Single-species algal cultures are now available at many petshops dealing in marine fishes and aquaria. You can buy a vial of the alga you think will look best in your tank and will do best with the other inhabitants and chemistry of the tank. Many cultures are rather non-descript in the vials, but they grow into luxurious carpets and filaments after they become established in the aquarium. Some typical cultures include such types as: a) *Cladophora* sp.; b) *Struvea anastomosans;* c) *Cladophora delicatula;* d) *Microdictyon boergesenia;* e) *Valonia utricularis;* f) *Ulva fasciata;* g) *Boodlea composita.*

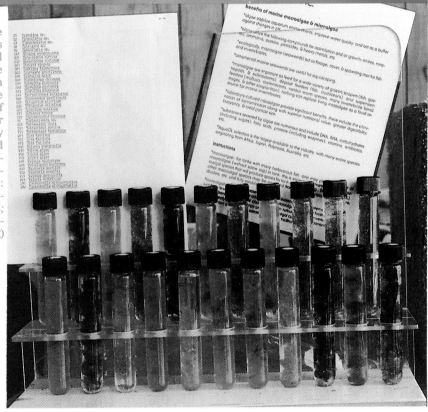

The large *Stoichactis, Radianthus,* and other tropical types are safe in suitably sized aquaria and are often displayed in association with anemonefishes. I have found *Stoichactis* and *Discosoma* anemones tolerant of lower light intensities than other genera, seemingly because their symbiotic algae flourish under less intense illumination than is required by the others, and they remain colored and healthy.

Small crabs, especially anemone crabs of various species and small hermit crabs, and various small shrimp are all acceptable members of a "natural" community. They will feed to some extent on living rock and may pick at sabellid or serpulid tubeworms, usually unsuccessfully, and are little nuisance if kept well fed with scraps. Avoid really predatory crustaceans like the mantis shrimps, and don't try to keep starfishes with harlequin shrimp (*Hymenocera* species), as the latter feed on them and also attack coelenterates. Large crabs, hermit or otherwise, should be avoided as they attack the scenery too enthusiastically. Immature lobsters and their relatives can be very attractive but are a considerable nuisance as they grow up.

Sea hares, although in general less attractive than nudibranchs,

Large anemones commonly used with anemonefishes are probably the largest invertebrates found in most marine aquaria. Although they are relatively safe and hardy, they do present the danger that if they should die they will rapidly pollute the entire tank and may result in a total loss of all life in the aquarium. Sponges are even worse in this regard than anemones.

Only recently have cultured giant clams, *Tridacna*, become available for the home aquarium. Small specimens of these unusual animals are very attractive, the often colorful mantle having a variegated pattern due to the presence of symbiotic algae. Clear windows in the mantle allow the entry of light for the algae.

feed on algae and so can maintain themselves, but once more, stick to small specimens. Quite possibly some will hatch out from the living rock. Nudibranchs are carnivores and mostly very specialized feeders that will not find much to eat in an aquarium. They will live for a few weeks or even months but eventually shrink and die. They also have an unfortunate habit of falling into anemones with horrendous consequences, both the nudibranch and the anemone producing slimes and toxins that kill them and other inhabitants of the tank or severely upset them. So, attractive as most of the nudibranchs may be, take care! If you decide to give them a try,

this is best done in a separate aquarium, where you can try feeding them on hydroids, sponges, and small encrusting invertebrates.

Most of the other small snail-like mollusks are acceptable guests, and some will emerge from living rock. Don't allow the aquarium to become infested with them or the algae may suffer too much. Some are also predators on other invertebrates, including other mollusks. Many small whelks are predatory, and some can even catch fishes. The cones (*Conus* species) are deadly, having a siphon that is used to shoot darts into unsuspecting prey. Once the dart has penetrated, the poison is pumped

Although very colorful, the flamingo tongue, *Cyphoma gibbosa*, is a mollusk that cannot be kept for long in the average aquarium. They feed almost exclusively on the polyps of gorgonians, animals that are seldom available in suitable quantity to be used as food.

through it from the cone shell's pharynx and can even cause death in humans.

Starfishes are a variable bunch. Most are liable to attack mollusks of any kind that allow them to do so, bivalves being particularly vulnerable. It is a good idea to keep some mussels in the aquarium to tempt the starfishes, as they are a readily accepted prey. A few echinoderms are algae-eaters or filter-feeders or feed on detritus, especially the featherstars and brittlestars. These are very likely to hide away under the scenery and all you see is an arm or two waving around at feeding time to gather up any tidbits. Various sea urchins are suitable if given a sandy base, as they are debris-feeders and need something to sift, encrusting algae also suit some species. Be careful in handling them as some are mildly poisonous and many can inflict nasty wounds that are slow to heal.

Suitable Fishes

Fishes suited to the natural system aquarium are those that do not make a specialty of eating algae, coral, or crustaceans. Many fishes will pick at these now and then but won't do too much damage and thus can be tolerated. Surgeons will denude a tank of your favorite algae in a few days, and angels will not be far behind. Young angels may do little damage, as they are carnivores and feed happily on normal fish foods, but as they grow up they are going to attack higher algae and even corals. Sponges don't usually flourish, but if they are doing so an angel will soon clean them up. Coral will be attacked by triggers, filefishes, and parrotfishes, while most wrasses and squirrelfishes will eat any small crustaceans they can get at in no time at all.

In nature the dense mats of *Halimeda opuntia* form a distinctive habitat in which many types of animals take cover and search for food. This is a coralline (calcium-depositing) green alga.

The green algae provide a large number of often hardy and bizarre types that can be grown in the aquarium. The usual shape and bright color of *Halicoryne* would make it an excellent aquarium alga.

Eels are far too predatory for most other fishy companions of small size. With butterflyfishes you have to be choosy—some will leave coral alone or not molest it too much, others devastate it. Enquire carefully about any you feel like trying out, and then be very observant once you put one into the tank.

So what does this leave? Plenty of attractive fishes that will live happily without severely disturbing the plants or invertebrates. First and foremost are the anemonefishes, best provided with suitable anemones. They won't devastate the algae or anything else as long as they get enough to eat, either naturally

produced plankton or added food. Other damsels are similarly acceptable as long as you avoid the few that, like the three-spot (*Dascyllus trimaculatus*), get large and very aggressive. Seahorses and pipefishes are ideal inhabitants if there are no other quick eaters around, but will not flourish in the presence of anemone fishes and other damsels. They can be accompanied by mandarins, except that mandarins, although also slow feeders, are very thorough eliminators of small invertebrates of many kinds. One mandarin in a large tank is enough. Gobies and blennies are usually suitable, too, many feeding on detritus, but avoid

The wisdom of this grouping of animals for the aquarium would be severely questioned by many advanced aquarists. The large anemones could produce enough mucus and waste products to endanger the more delicate animals in the tank, while the predatory nature of starfishes is well known.

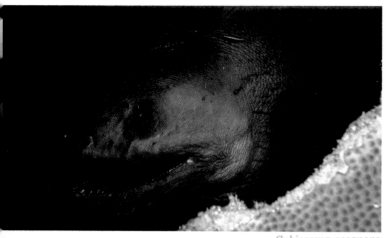

Gobiosoma oceanops

Gobiosoma illecebrosum

Gobiosoma randalli

Gobiosoma evelynae

Gobiosoma xanthiprora

Gobiosoma genie

Chaetodon triangulum

Chaetodon larvatus

Chaetodon leucopleura

Chaetodon trifascialis

Chaetodon paucifasciatus

Chaetodon madagascariensis

Chaetodon melannotus

Chaetodon ocellicaudus

large and possibly predatory species. Boxfishes, if small, are also a possibility.

There remain some of the larger fishes, mainly predators that will not worry about small crustaceans or fishes they cannot swallow. Lionfishes and anglers, properly fed of course, live quite harmlessly in a natural tank, but the lions *will* clean up any suitable crabs and shrimp, depending on the size of the fish and of the potential prey. Decorative eels love a natural tank with its hidey-holes and can be kept with other large fishes they cannot eat. On the whole, though, there is better scope for the creation of an attractive scene with the smaller, non-predatory species mentioned above, as far as the fishlife is concerned. Keep the numbers down to give the invertebrates a fair chance.

Miniature Reef Aquaria

As stated earlier, this book is not going to discuss the "minireef" type of aquaria at any length, but it should be pointed out that they are evolutionary developments of the natural system. They bolster it up so that really crowded tanks can be maintained with little trouble and almost no danger of a wipe-out, as long as they have been properly set up and maintained. They allow the culture of reef-building corals to a greater extent than the natural system, particularly those with intense lighting. But for this step forward in marine aquarium keeping you pay heavily pocketwise and the really advanced versions cost thousands. It's a bit like the difference between a Ford and a Maserati; both do a good job and you pay a great deal for the refinements of the more expensive car. If you are really keen and can afford it—why not?

Chrysiptera flavipinnis

Chrysiptera cyanea

Chrysiptera taupou

Damselfishes are colorful, hardy, diverse, and readily available. Often they can be spawned successfully in the aquarium. They are very territorial, however.

A diagrammatic rendition of internal sexual differences in fishes, using *Platax orbicularis* as the model. In male fishes (right) the testes are closely connected to the kidneys and often open through the same duct. In females of many fishes (left) the ovary is a large sac-like structure that releases the eggs directly into the body cavity and collected by an ovarian funnel to be transmitted outside.

In aquarium fishes, there are normally separate ovaries and testes producing eggs and sperm respectively. The ovaries shed their eggs into oviducts as in other vertebrates and the testes shed sperm into ducts derived from part of the kidneys that in turn lead either to the ureters or to a separate spermatic duct. The hormonal control of the gonads (ovaries and testes) is similar to that in the higher vertebrates and involves pituitary gland hormones, without which the gonads do not function. The gonads in turn produce male and female hormones that determine the outward sex and behavior of their possessors. This, however, is not the whole story, as some marine fishes change sex or are held in a juvenile state by the presence of mature adults, while in rare instances both ovarian and testicular functions may be present in an ovotestis in the one individual. In still rarer instances the sperm may fertilize eggs from the same ovotestis, as in the Belted Sandfish, *Serranus subligarius*, of the Florida reefs.

To the aquarist, the most important fishes at present that undergo sex changes are the anemonefishes, because they can be bred in the aquarium. In some species the usual group of a mated adult pair and several immatures living in one anemone exhibits both persistent immaturity in the presence of adults and protandrous sex change (from male to female) if the female dies. Then the male becomes a female and one of the immatures takes his place. All done by mirrors! Actually, by pheromones—chemical substances released into the water by the adults that keep the male a male and the immatures immature until one of them, presumably the most advanced, is needed as a replacement. Similar changes have been reported in other damselfishes and wrasses, the latter being protogynous in some species (changing from female to male).

The stimulus to the pituitary gland to start the gonads going differs in different species, often depending on changes in the length of the day, in temperature, and in water conditions. In the North Sea, most fishes spawn in spring and early summer as both daylight increases in length and the temperature rises, but such changes are insufficient in the tropics, where often no sharply defined breeding season is found. However, some fishes breed in association with monsoons, triggered perhaps by dilution of the water by heavy rains. What determines a breeding season in

It is being found that many marine fishes, especially groupers, are hermaphrodites, with both sexes in the same individual. In extreme cases (which are quite rare) the sperm from one animal may fertilize the eggs of the same animal. This is what happens in the Belted Sandfish, *Serranus subligarius*.

deep-sea fishes remains unknown, yet larval anglerfishes spawned by adults that never rise above 3,000 ft are found in surface waters at definite seasons.

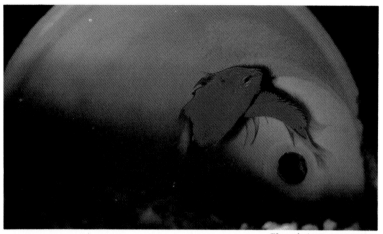

Chrysiptera parasema

Chrysiptera rollandi

The adults live in the dark, apart from luminescence from themselves and other creatures, at a temperature of 35°-40°F. The famous example of the grunion (*Leuresthes tenius*), which spawns only following a full or new moon, is still a mystery, but a guess is that the fish is sensitive to the tides, not to the moon itself.

Marine fishes very rarely care for their young. Sometimes they look after their eggs until hatching, but most often they produce floating eggs that become part of the zooplankton. This does not prevent attempts to breed them in captivity, but it doesn't make it easy. Most newly hatched fry cannot take food as large as can many freshwater fishes, and this compounds the difficulties. It isn't particularly hard to get a variety of marines to mate and lay fertile eggs; raising the young is the main problem.

Quite a number of species of interest to the aquarist live on reefs or near the shore and lay their eggs on the bottom or on rocks, shells, or weeds, even in a nest of some sort and are known as demersal fishes. Some that do this and guard the eggs are found among the gobies, blennies, damselfishes including anemonefishes, sticklebacks, and basslets. These are naturally the species on which most attention has so far been focussed. Some help has been given, however, by the experiments of hatchery scientists with such fishes as cod, haddock, flounder, and plaice. It comes as a surprise to many that as early as 1917, the total output of fry from the three American East Coast hatcheries was over 3 billion per annum. The work began in 1878 on the cod, eggs and milt being obtained by stripping fishes. Various other countries took up the story, and soon hatcheries were operating all over the world and producing many millions of fry for release into the ocean. By the 1950's it was practically all over, as it could not be convincingly shown that the release of fry resulted in increased catches, and the practice was deemed to be uneconomical. The value of the work to present-day scientists and aquarists lies in the findings relating to the rearing of fry, although regrettably the newly hatched brine shrimp that so much aided the old-timers is too big for most of the aquarium species.

Obtaining Breeding Pairs

It is not often possible to purchase breeding pairs of marine fishes. As it is usually difficult to sex individual fishes, the best way to obtain at least one pair is to purchase a small group either of presumed adults or of young fish to grow out. Statistically, a group of six fish gives about the least acceptable chance of not obtaining a pair; this chance is one in 32. Whether you can afford six fish, young or old, depends on the species and yourself! The only other way is to observe groups of the desired species in the dealer's tanks and to see if you can make a successful guess. The female is often fatter and larger than the male, deeper in the belly even if not full of roe. The male tends to be more colorful than the female and more belligerent in some species, but only in adults. In the seahorses and pipefishes the male often has a typical brood pouch on the abdomen. In some of the damsels, especially the anemonefishes, a small group may be expected to produce a pair, since they pass from male to female or immature to male "on demand". Presumably the largest fish in a group becomes the female, followed by the next largest, who remains male, so as long as they survive only a few fish should suffice. I have not found this question clearly discussed.

It is a pity that some marine fishes do not readily tolerate members of the same species and that in general it is usual to have only one specimen of each per tank. This makes the chance finding of mated pairs less probable than it is with freshwater fishes. It must still be a more frequent occurrence than deliberately looking for them or planning for them, but it is clearly not often followed up. The pair mate in a community tank and that is the end of the story. I have accidentally witnessed in my own tanks the mating of seahorses (*Hippocampus novae hollandiae*) and raised their young; of anemonefishes (*Amphiprion frenatus* and *A. ocellaris*) and raised some of *A. frenatus;* and the spawning only of anglerfishes (*Antennarius striatus*) whose young defied attempts to raise them; and the spawning of mandarins (*Synchiropus picturatus*) and of the blue damselfish *Pomacentrus coelestis*. In the last two instances I didn't even try to raise any young, although the mandarins spawned repeatedly. In a large community tank I have at present, an interspecific pair composed of a male *A. frenatus*

and a female *A. ephippium* has spawned well over 100 times in the course of six years, producing an estimated 30-40,000 larvae, not one of which has survived.

This is exactly to be expected!

Once an adult pair has been found, it should be removed to a separate tank for breeding and be given suitable conditions. This prevents stress and gives the fish an optimum chance of producing fertile eggs not immediately consumed by onlookers, although in most cases they will consume them themselves. However, some species spawn best in groups and are to be avoided. That is why there are very few accounts of the successful breeding of damsels (other than anemonefishes), wrasses, and ocean fishes.

Bottom-dwelling fishes that place their eggs on a rock or shell and look after them are, next to seahorses, the easiest to breed, and the easiest of these are various anemonefishes. If you want to have a go at breeding any marine fish just for the experience, these are the ones to choose.

Breeding Tanks

In contrast to freshwater fishes, marines are usually left in a more or less permanent breeding aquarium and the larvae are removed, not the parents. The breeding tank is therefore set up just the same as an ordinary

Amphiprion mccullochi   *Amphiprion melanopus*

Amphiprion melanopus   *Amphiprion melanopus*

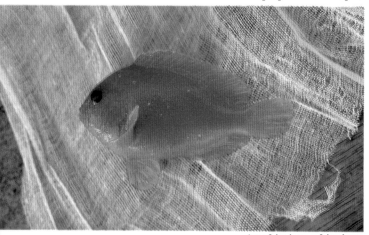

Amphiprion ephippium   *Amphiprion ephippium*

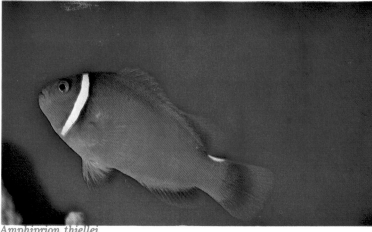

Amphiprion bicinctus   *Amphiprion thiellei*

Amphiprion sandaracinos   *Amphiprion perideraion*

Amphiprion tricinctus   *Amphiprion chrysopterus*

Amphiprion rubrocinctus   *Amphiprion rubrocinctus*

Amphiprion frenatus   *Amphiprion frenatus*

The lionfishes or turkeyfishes, genera *Pterois* and *Dendrochirus*, have long been the favorites of many marine aquarists. These large, colorful fishes have distinct personalities, plus they have the additional "mystique" of being venomous, the spines bearing poison glands that can cause painful (rarely even deadly) stings. Although they have large mouths, they seldom prey on the larger fishes in the aquarium with them. This is *Pterois volitans*.

aquarium and serviced in the same way. It should be larger than theoretically necessary for a pair of the particular species being bred, to give room for courtship, to give optimal conditions, and to facilitate removal of the larvae. A 20-gallon tank is about the smallest useful size for bottom-dwelling fishes, but even small angelfishes need twice that size and larger ones up to 60 or 80 gallons. They must have room to swim around and release their eggs into the water where they form part of the plankton in nature. Anglerfishes are an exception because they produce rolls of floating jelly containing the eggs and will spawn in a 10-gallon tank. Also exceptional are seahorses and

pipefishes, as the male carries the young in his pouch, and mouthbrooders like the cardinalfishes. Anemonefishes behave more naturally and spawn more readily at the base of a suitable anemone, which should be provided if possible.

Marine fishes from the tropics usually do not need any change of conditions to induce spawning—again in contrast to many tropical freshwater fishes. This is as long as a natural lighting period and temperature are adhered to, say 14-10 to 12-12 hours light-dark and about 78°-82°F. Some accounts state that a lowered salinity has induced spawning in bottom-dwelling fishes, from 1.022 to 1.016 or 1.026 to 1.019 for example, even overnight,

which sounds a little heroic. It's worth a try if nothing happens after waiting a while, but I think I would do it in two or three stages over a period of several days. Then what to do? Raise the larvae at the lower salinity or put it up again before they hatch? Your guess is as good as mine.

When they spawn, bottom-dwelling (demersal) fishes typically will produce a few hundred eggs laid on a cleaned area, in a nest or sometimes just scattered on the bottom. Small open-water (pelagic) fishes produce similar numbers, but per day, with totals in the many thousands, whereas large pelagic fishes can run into tens or hundreds of thousands per day. The demersal eggs have hatching

---

Anemonefishes are somewhat unusual hermaphrodites because males develop from females. When the male of a pair or group dies or is removed, the dominant female changes sex over the course of a few days or weeks, taking his place. An immature matures into a female. This seems to be the rule for all species studied so far. Shown is spawning in *Amphiprion allardi*.

The bright yellow or red egg masses of anemonefishes are becoming familiar sights in aquaria. Many species have now spawned in captivity, although it is still difficult to rear the young through the larval stage. Shown is *Amphiprion clarkii*.

periods of up to a couple of weeks, depending on temperature and species, whereas the pelagic eggs hatch within a day. If they didn't, even their vast numbers would not provide enough eventual young fishes. Mouthbrooders' eggs have a long hatching period, like the demersals, and fewer still are produced for obvious reasons. The seahorse carries his young for six to seven weeks and produces a few dozen to a few hundred, according to species and the size of the male. Herein lies an interesting story. You can produce a dwarf strain of the common Australian seahorse *Hippocampus novaehollandiae* by feeding its young only on newly hatched brine shrimp. They grow to about 3″, instead of the 5″ or 6″ of the parents, and breed at that size and produce only a few dozen young at a time instead of the usual several hundred. Their young repeat the process, but no doubt if properly fed they would grow to the usual size and

Male seahorses giving birth to their young are a common sight today, but the event never fails to draw attention. Few seahorses are actually bred in captivity, but wild-caught specimens often are pregnant.

produce the usual numbers of young, as their genes won't have altered.

Spawning Behavior

Although a typical coral fish lives in a restricted area, often not wandering more than a few feet from "home", many of them are pelagic spawners and resemble some of the egg-scattering freshwater fishes in general behavior. Not many, however, have been carefully observed and we may be in for some surprises. Some pair off semi-permanently, unusual in egg-scatterers, as does the Harlequin Bass (*Serranus tigrinus*), which pairs off and shares a territory with a mate, although both are hermaphroditic (possessing both sexes at once). The mating procedure involves a rush upwards through the water above the reef, when eggs and

milt are shed after a brief courting display with fin flicking and curling motions of the body. Spawning, as with many reef fishes, is usually at dusk, shortly before the sun sets. The Jackknifefish (*Equetus lanceolatus*) also pairs up but is not hermaphroditic, and also schools on occasion, but whether for spawning is not clear. Some angelfishes and some butterflyfishes are observed to pair off in nature, seemingly for long periods, so the pairing habit is by no means rare in reef fishes that are pelagic spawners and take no interest in the eggs once they are released.

Both Pacific and Atlantic anglerfish spawnings have been observed, with some indications of differences. The male Atlantic angler *Antennarius scaber* has been described as guarding the female when she was blown up with

spawn, whereas my own observations of several Pacific angler (*Antennarius striatus*) spawnings have shown no such behavior. Instead, no notice appeared to be taken of one sex by the other until a brief chase around the tank was followed by release of the roll of jelly, containing many thousands of eggs, that floated on the surface and disintegrated after a few days, by which time developing embryos were clearly visible and hatched in four to five days from spawning. The Atlantic angler eggs appear to follow the same pattern.

Mandarinfishes, both *Synchiropus splendidus* and *S. picturatus*, are, rather surprisingly in view of their strictly demersal habits in normal circumstances, pelagic spawners. They also give a most spectacular performance. The male chases the female

The Harlequin Bass, *Serranus tigrinus*, is unusual among egg-scatterers in that it forms stable reproductive pairs. Although both members of a pair are theoretically able to produce both eggs and sperm, they instead usually act as though they were single sexes when spawning.

Although they are bottom-dwellers, the mandarinfishes (including *Synchiropus splendidus*, top, and *S. picturatus*, bottom) court and lay the eggs in midwater.

Antennarius striatus is a frogfish in which the breeding habits are now fairly well known.

The thousands of small eggs are laid embedded in a gelatinous "scroll" that does not disintegrate until the eggs are fully developed. The scroll is much larger than the fish when it absorbs water from the surroundings.

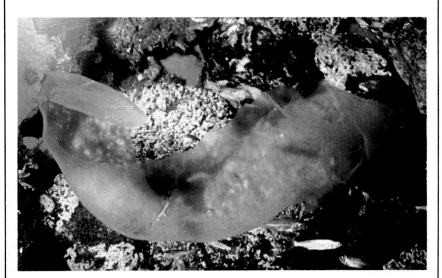

Detail of the scroll of eggs removed from the water.

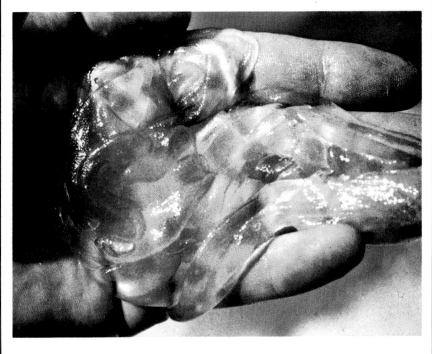

Antennarius striatus (left) is a typical anglerfish having the front dorsal spines modified into a lure and bait to attract prey.

around the tank and displays with erect fins and several circling motions above her, while she remains at rest. Later he chases her up into the water. After several such ascents they spawn side by side and then return to the bottom of the tank. Despite pelagic spawning, only a few hundred eggs are released.

There have been more accounts of demersal spawnings than of pelagic ones, of the anemonefishes in particular. Various *Amphiprion* species behave very similarly, forming permanent pairs with their cortege of immatures quite

frequently present, but not always and not, of course, in spawning tanks where they are separated. *Premnas biaculeatus*, the Maroon Anemonefish, behaves similarly to the *Amphiprion*. As an example I shall take the Tomato Anemonefish (I refuse to call these beautiful fishes "clowns"), *A. frenatus*.

Amphiprion frenatus is rather more belligerent than most other species and gives a good display of territoriality at every chance, whether having spawned or not. The pair chooses a site under an anemone if possible and guards it

against all comers, although in a populated aquarium they learn to tolerate the presence of many of their tankmates, but rarely of other anemonefishes of any species. As the female blows out with spawn, they both clean an area of rock or coral by biting away at it for several days. The male attacks most other fishes, or the aquarist, and can draw blood. The female makes aggressive

A view of the unexpanded scroll of *Antennarius pictus* before absorbing much water.

shuddering movements, even through the glass at someone outside the tank, but does not often complete an attack. A certain degree of courtship takes place, with similar "shimmies", while the cleaning is going on. Spawning occurs late in the afternoon and is completed in one or two hours. The female lays adhesive orange-colored eggs in closely spaced rows on the cleaned area, but often overlapping it, and the male follows with his milt at each pass. Both guard the eggs and mouth them frequently, probably displacing by accident any that fail to develop. Usually they all develop, as far as can be observed.

An account of the spawning of the Yellow-tailed Damselfish, *Microspathodon chrysurus*, regrettably did not describe spawning behavior in any detail or even state whether one male is mated to several females as might be expected. It seems so, since a nest of eggs covers up to 80 square inches of coral and contains up to 92,000 translucent, pinkish eggs. These could hardly all come from one female. The eggs are apparently in an exposed position and hatch in about three days at 80°F. My own observations of the spawning of the Demoiselle, *Pomacentrus coelestis*, were limited by the fact that the male dug out a nest in the substrate and enticed the three available females into it, where they deposited invisible eggs that he guarded very fiercely until, presumably, they hatched out at night and were eaten. This also took three days, of guarding anyway. To get the females into the nest he showed off very brilliantly, glowing with color and exhibiting the usual shaking of the head and body, swimming down into the nest and being

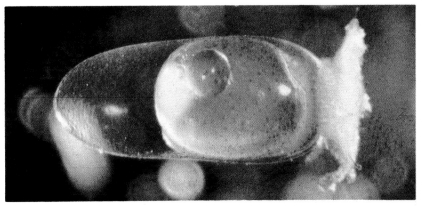
Development of *Amphiprion chrysopterus*. Egg at 36 hours.

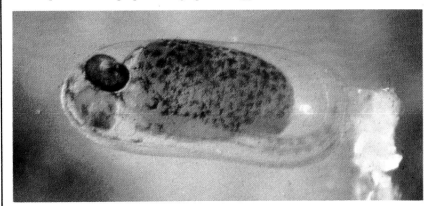
104 hours.

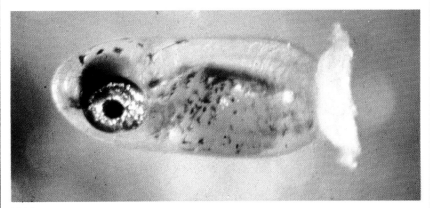
126 hours.

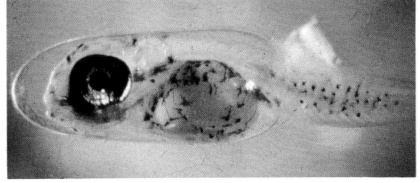
Well-developed larva in 148-hour egg.

followed by the female. If she didn't respond, he would sally forth and give another exhibition, rarely chasing her or showing the rather aggressive behavior more characteristic of anemonefishes. Another male in the tank was chased off whenever he approached but was not harmed, as I imagine he would have been in a smaller tank, as this one was a 6-footer with plenty of hiding

chewing at it as if to clean or aerate it. The mass contains 150 to 300 yellowish eggs attached by fibers to a "mainstem" and forming a spherical body like a bunch of grapes. The eggs are quite large, a little over ⅕ inch in diameter. The period of incubation was not given, and of more than 12 spawnings apparently only one was raised.

The aquatic "dance" of

seahorses of various species has often been described, with the female inserting her eggs into the pouch of the male. Eventually he shoots the hatched young into the water by convulsive movements, where they take up an independent existence and are not attacked by the parents or other seahorses and so can be left in with them.

places. The Blue Devil, *Abudefduf cyanea*, has been described as spawning in pairs, depositing only 15-45 eggs.

An example of a mouth brooder is the Yellowhead Jawfish, *Opistognathus aurifrons*. The fish lives in individual burrows in small colonies bordering coral reefs. The eggs are kept in the male's mouth almost constantly but are deposited in his tunnel for a short time while he feeds. Most of the time he just sits at the entrance to his tunnel quietly holding the egg mass, but from time to time he spits it out and recovers it,

In some damselfishes the brightest colors are present only during mating displays or while guarding eggs. Above is a spawning male *Chromis viridis* and the less colorful females; below is a gorgeous *Chrysiptera starcki*.

Collecting Eggs and Larvae

With the demersal fishes that care for their eggs it is best to leave them with the parents until hatching. If you can see the developing eggs, not only the incubation period typical of the species but also the stage of development will enable prediction of a likely hatching date. If convenient, you can remove the eggs while they are still attached to their base and transfer them to a bare tank containing some of the same water. They will hatch out during the following night, sometimes over two nights, but usually in a batch an hour or two after dark. Of the most commonly bred genera, *Dascyllus* eggs hatch after two days and *Amphiprion* eggs

after eight days at the typical 78°-80°F tank temperature, but can take up to ten days at lower temperatures. The eggs range from orange to brown and show the embryonic eye spots a few days before hatching, with the larvae sticking out from the base and waving around on stalks as the parents fan them. When the eyes become gleaming and prominent, look for hatching to occur.

As a probably better alternative, wait for the larvae to hatch out in the parents' tank, say two or three hours after dark, and then shine a flashlight at a top corner of the tank. The now pelagic larvae are phototropic and will swim up to the illuminated area and concentrate there.

Other than a few cardinalfishes, about the only mouth brooders kept in the marine aquarium are jawfishes. The Yellowhead Jawfish, *Opistognathus aurifrons*, of the Caribbean will breed in the aquarium if given the proper substrate in which to tunnel.

Enough of them for rearing can be caught either very gently by a siphon or baster or equally gently with a small vessel of some kind, *not with a net,* which would damage them. Pelagic eggs and larvae can be collected from the surface in a similar fashion, except that the eggs won't swim toward a light. There are so many eggs that you will be able to collect enough, just a few hundred, without much trouble. It is probably best to collect the eggs and not the larvae.

The Rearing Tank

The bare rearing tank, applicable to all types of breeders, need be only of a suitable size and reasonably deep. The bigger it is the better, within reason, but as it will require rather large water changes it is best to consider the availability of these before adventuring on to too large a tank. Something like 10-20 gallons per 100 eggs or larvae is needed, so tailor the numbers you try to raise to the size of the tank you can provide and service. There will need to be only gentle aeration, provided by about one stone per 20 gallons giving as fine bubbles as possible, and good top lighting to keep the larvae well up in the water. The only other requirements are heating and a thermometer. The temperature should be about the same as it was in the spawning tank and should not vary much. The pH should be around 8.0 and should also be kept as constant as possible—the water changes will probably take good care of that. For the same reason, do not worry about ammonia or nitrite levels as the larval biomass is very small and should give no trouble unless you overfeed grossly and do not siphon it off.

The larvae can only eat when the light is on, so a longer period of lighting than usual seems indicated, despite the fact this does not happen in nature. Although freshwater fish breeders sometimes raise larvae in constant illumination, I have never seen this advocated for marine larvae, nor have I tried it. It should be tried, as the results with freshwater fishes are accelerated growth and no obvious harm. The only trouble would be maintaining a suitable food level, but this is not beyond the capability of the serious breeder. A constant drip device and a mechanical feeder of some type would be possibilities. The advantage of a dark period in nature is to the enemies of the young larvae, many of which rise up from deeper waters and can feed on them, together with other tiny zooplankton, without having to see them.

As the larvae remain pelagic for several weeks and the top light keeps them up in the water, cleaning is easy and can be done by siphoning from the bottom where the waste accumulates. It is best done every few days at the total rate of about 10% per day if the other recommendations given are followed. The longer you leave a change the bigger it has to be and the more dangerous it may become. Frequent small changes are much the best, made with the least disturbance possible. The new water must be properly conditioned and be as similar to that in the tank as possible—same mix or same sea water batch, same temperature and pH, adjusted with much greater care than you might normally take. Larvae are fragile individuals and will repay gentle, thoughtful handling.

After a few weeks in a typical species of demersal fish, the growing larvae, now really young fish, will desert the plankton and settle on the bottom. At this stage they should be moved to a normally furnished and undergravel-filtered tank—perhaps back to the tank in which they were spawned, if their

For best results when attempting to rear fry, the separate rearing tank should be bare except for a sponge filter or equivalent and provisions for heat. Longer light periods than usual may be necessary.

parents are no longer in it. Even pelagic species need such a tank for best growth, as they are becoming big enough to need the biological filter. By this time all types will be feeding on a varied diet and there will be no need to fiddle with frequent water changes and first planktonic foods. As the young fishes grow they must be given adequate living space by transferring them to several tanks or by disposing of some of them.

Feeding Larvae

Newly hatched larvae feed on the contents of the yolk sac for a day or two, but at the latest by the end of day three they need feeding and will die if not fed promptly. This means being ready on that day with a suitable first food. This is not newly hatched brine shrimp, which are too large for most species of tropical marine aquarium fishes. Instead, some form of minute organism has to be cultured in large numbers if natural plankton is not available, and it usually isn't. This has been the nemesis of the great majority of amateur attempts to breed marines, and many of the professional ones as well. Attempts to substitute artificial diets, often quite useful with freshwater species, have rarely been successful. All sorts of diets have been tried—finely divided egg yolk, protein suspensions, ground or minced up fish foods of many kinds, yeast, even blood cells, but none has worked satisfactorily except frozen rotifers with anemonefishes. As freshwater rotifers are easy to breed, it seems odd that nobody appears to have tried them. They should float in seawater and could well be consumed even after death if frozen marine rotifers are taken.

Although marine plankton would seem to be a great solution if available and is advocated by various authors, it isn't an unmixed blessing. Tropical plankton lives in the aquarium long enough to be eaten, but that from colder waters does not. Both types may contain parasites that can play havoc in the rearing tank. The majority of aquarists, who cannot get plankton of any sort, need not bemoan their fate too much as it is safer and in the end easier to culture planktonic organisms foods for yourself. It is zooplankton that primarily concerns us, but zooplankton have to feed on something. Even if they are carnivorous it eventually boils down in nature to phytoplankton and in artificial culture to that or a suitable substitute. Phytoplankton is the basis of all food chains in the sea and consists mainly of unicellular algae that derive their substance from the sunlight, the water and salts of the ocean, and the gases of the atmosphere. Zooplankton of all types, whether fish eggs and larvae, minute crustaceans, the early stages of myriads of species of sea creatures, or the larger euphasid type shrimp and other crustaceans all feed on phytoplankton or each other.

It is easy to start a freshwater infusorial culture to feed newly hatched freshwater larvae, but boiling up some plant material in sea water and exposing it to the air won't give the same result. It just grows bacteria, as marine planktonic organisms are not conveniently floating around in the air to start up a culture as are freshwater ones. Come to that, even the freshwater culture goes better with an inoculation of living organisms to start it off. We therefore have to start up a marine culture either with phytoplankton or a substitute for it and then inoculate it with a desirable species of zooplankton and hope to keep it going. We

can now purchase both phytoplankton and zooplankton cultures from many dealers.

Algal Cultures

There are several ways of obtaining sufficient algae to feed the zooplankton that is required for the larvae, as they do not eat phytoplankton direct. A rough and ready method is to get some marine water from an established aquarium and inoculate a few gallons of fertilized sea water with it, keeping it well illuminated. Constant lighting is quite in order. By this method unselected species of algae will grow and give you a medium in which the zooplankton should flourish. Water from an aquarium, other than a natural system one, will not contain unwanted and probably unsuitable zooplankton together with the algae so a phytoplankton culture is obtained. Suitable fertilizers for marine algae are sold in pet shops, but in a pinch a little indoor plant fertilizer that will contain the necessary elements can be used. Use about one hundredth of the concentration recommended for watering house plants—in other words, use a liquid preparation as made up for watering at the rate of about 1 oz per gallon of sea water. Double this amount per day for the first few days would not be wrong, but once the green water that results is good and strong, cut it down to the recommended dosage. A good culture has unbelievable numbers of algae—about 100 million cells per teaspoon!

It is much easier to inoculate the fertilized water with a known algal species. These can be purchased commercially. Algae such as *Chlamydomonas, Anacystis, Dunaliella,* and *Isochrysis* may be used, but the present vogue is for *Chlorella* algae, easily obtainable in pure culture. These algae are a sufficient food for many zooplankton species, but there is no objection to supplementing them with dried baker's or brewer's yeast in modest amounts or, perhaps better, with a marine yeast. Torula yeast is available from health stores and actually grows in sea water and has even been used on its own; so have freshwater yeasts with good success, particularly with rotifers. There seems to be a general feeling that yeast alone, of any kind, is probably an insufficient food, but experience shows that you *can* raise anemonefish young with rotifers fed on yeast.

Zooplankton Cultures

The first foods that appear to have been cultured for marine fish larvae were copepods, as early as 1915. The methods used were developed in the laboratory and, as might be expected, involved a fair amount of equipment, but they can be reduced to quite simple jars and containers sufficient in production for the aquarist. Brackish-water copepods of the genus *Acartia* were favorite subjects. These are calanoid copepods that swim constantly in the body of the water and can live in full strength marine water if they have to. The adults are too large for larval fish to eat; it is their nauplii that are used as food. These copepods are like the familiar *Cyclops* of freshwater pools. A marine copepod, *Cetochilus* is devoured in large quantities by whalebone whales. The female typically carries two egg sacs from which the eggs drop into the water and is so prolific that one individual, should all her eggs survive and produce young in turn and so on, could produce over four billion

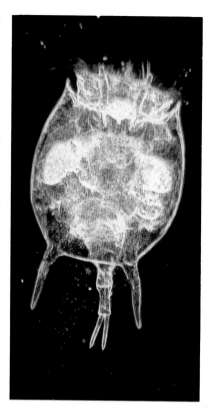

Currently the food of choice for larval marine fishes is rotifers, especially *Brachionus*.

young per annum. Harpacticoid copepods live at the surface of the water and could be cultured as are the calanoids since they have free-swimming nauplii. They do not appear to have been used, but there seems no reason why not.

To culture copepods, up to 100 females are placed in vessel with a fine screen through which the eggs drop into a larger, surrounding vessel. A 1/2-gallon inner and a 1-gallon outer vessel would be adequate. The eggs are siphoned off from the bottom of the outer vessel and placed in hatching jars of about 5-gallon capacity. Both the egg-laying females and the nauplii are fed with a rich algal suspension in sea water and are kept well illuminated to keep the algae flourishing. With all such cultures, it is necessary to keep more than one batch going, as they are subject to sudden

collapse and you might otherwise find yourself short of food for the larval fishes at a critical stage.

A favorite alternative to copepods is rotifers, of which many different species could be used, but the established one is *Brachionus plicatilis*, a brackish-water and marine species about half the size of a newly hatched brine shrimp. It can be purchased from dealers and is more easily cultured than are copepods. Rotifers are easily identified by the circlets of cirri on the head end that wave in the water and give the appearance of rotating wheels, hence the name "wheel-animalcules". Some are as small as 1/500 inch in length, and the largest is only about 1/8 inch. *B. plicatilis* adults are less than 1/100 of an inch long, the female being larger than the male. Males are rare, and most of the reproduction is by parthenogenesis, development from unfertilized eggs. In adverse conditions males are produced and resistant eggs are laid that can lie dormant for several years. This is a danger in cultures, which must be kept in a favorable state to prevent such a happening.

Rotifers are just as quick breeding as copepods, their eggs hatching in a day and their young growing to adulthood in another day, ready to reproduce. *Brachionus plicatilis* can stand large variations in salinity and temperature and is not demanding of great care as long as it gets plenty to eat. The optimum salinity for breeding is said to be 3% to 4% (i.e., ordinary sea water) and the optimum temperature 80°-90°F, a bit warmer than the average tank. It can be cultured in containers as described for copepods or even smaller ones down to 1/2 gallon size, but the larger the better for

Typical spawning antics of a pair of *Chromis atripectoralis*, a reef-dwelling damselfish. Most marine fishes from the reef have similar open-water display movements as courtship behavior.

stability and length of life of a culture. In this case the adult breeding females, not their eggs, are added to the culture together with generous supplies of one-celled algae. Any white fluorescent light will keep the algae healthy and growing, although more will have to be added from time to time as the rotifer population increases. It doesn't much matter how many rotifers you start with, as long as there are initially sufficient to dominate the scene—we don't want a useless intruder to swamp out the rotifers, and the only other way to prevent this would be sterile cultures of both algae and rotifers, a difficult and time-consuming procedure. Aeration at not too brisk a rate is advisable unless the vessel is very shallow. If you are using an ordinary tank to rear the rotifers, it is needed.

Once a satisfactory setup has been achieved, you will want to keep rotifer production moving at a mild pace until high production is needed. To do this, lower the temperature to not less than 60°F and feed a quarter as much as has

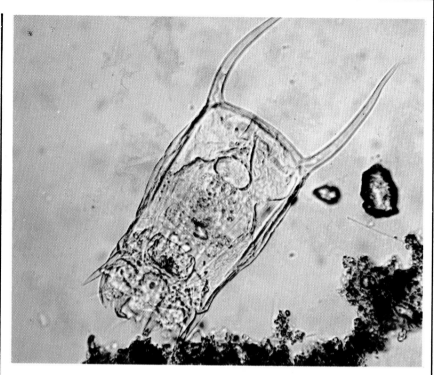

Although there are literally hundreds of species of freshwater rotifers available in ponds and other bodies of water, there are fewer common marine species. The marine rotifers are seldom easily collected even if you have access to the proper equipment and often do not do well in cultures. For this reason it is best to purchase a good breeding culture of *Brachionus* from a dealer.

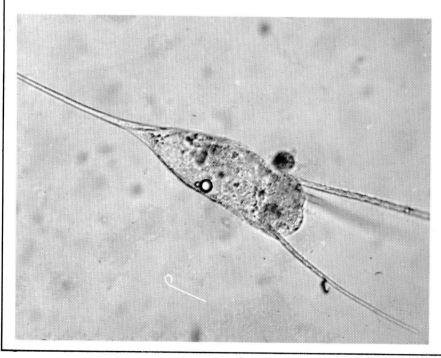

Marine rotifers are often more heavily ornamented with spines and projections (to help them float passively in the water) than are freshwater species. Whether the presence of long spines on a rotifer reduces its value as food for larval fishes is not certain but seems likely.

Copepods have long been favored foods with European hobbyists, who seem to have the knack of raising them in large numbers. They even feed the larval stages to young fishes. There is a danger, however, that large numbers of copepods in an aquarium will prey upon small fishes.

Daphnids such as these *Bosmina* are readily collected near shore but are often contaminated with larval tapeworms and other parasites. Pure cultures of marine daphnids are available from dealers, but they are usually considered inferior to rotifers as larval fish food.

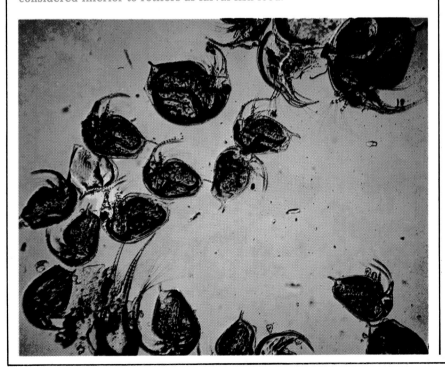

been found necessary at the higher production rate. When a need for feeding of fish larvae is anticipated, give the cultures a two-week start as they take some time to reach full production again. I say "cultures" because it is highly advisable to keep several going since, again as with copepods, a culture can fail or go bad for one reason or another. A density of around 20 rotifers per drop (2000 per standard 5 ml teaspoon) should be attainéd before their use as food. The adults can be harvested by filtering them off through a 40-micron mesh screen or portions of the culture water can be placed straight into the aquarium. The former method is best, as it avoids green water in the rearing tank that may become too dense for comfort and safety, although a mild degree of it is favorable rather than otherwise. Whichever technique is used, keep the culture vessels of algae and rotifers topped up so as not to lose productivity. Don't forget that rotifers can be raised on yeast if there is a problem of algal supply—use just enough to keep

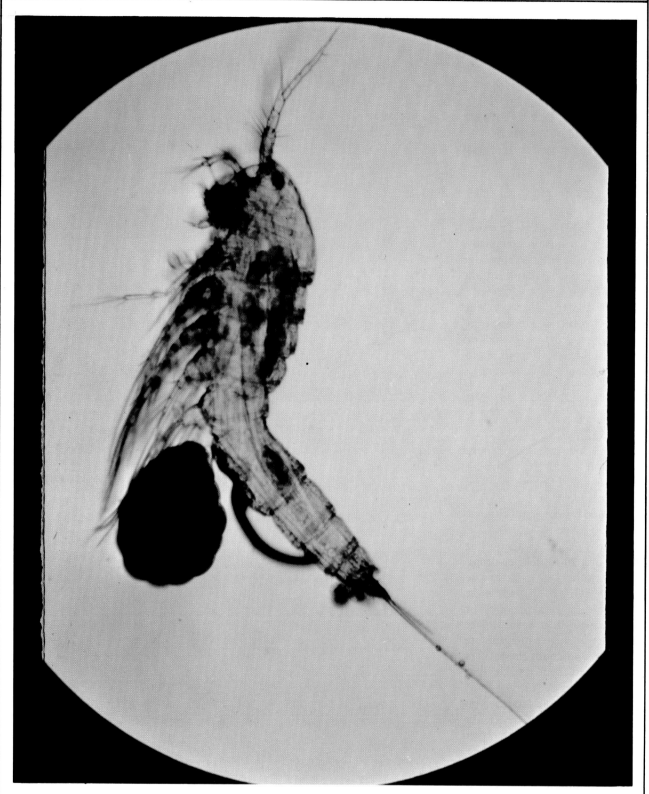

Many of the copepods collected in any locality will be found to be females carrying large egg sacs under the abdomen. The larvae from these eggs are fine larval fish food as they seldom carry parasites and are the proper size for many species. Copepod larvae will under certain circumstances kill very small fishes, so be careful.

the culture water mildly milky in appearance. Uneaten yeast will accumulate on the bottom and can be siphoned off gently.

There is now good evidence that juveniles raised only on rotifers may not be too robust, and that feeding on copepod nauplii is to be preferred, at least in part.

Feeding the Fish Larvae

The next consideration is how much to feed. The answer quite obviously depends on how many larvae are to be fed and how large is the tank holding them. But the obvious is not always correct. The true answer is that a certain density of rotifers (or larval copepods) must be maintained. A larval fish cannot be expected to thrive if it has to swim several inches between gulps—the food must be all around it, ready to be eaten. One larval fish in a 20-gallon tank needs as many rotifers as 100. Experience shows that 5-10 rotifers or any other similar foodstuff per ml is adequate and should be maintained by several additions per day. This is 20-40,000 rotifers per gallon, or about 3 oz (90 ml) of a 20-rotifer-per-drop culture, strained off or otherwise.

To determine the rotifer count of the culture, place one drop on a microscope slide, warm it over a flame or whatever to evaporate the water, and count the rotifers. You can stain them after drying with methylene blue if you find difficulty in counting when using a good hand lens or a low power of a microscope. Check that your "drop" is reasonably near to the standard 0.05 ml (20 drops per ml) by dropping 20 drops into a 1 ml measure or 25 drops into a ¼ teaspoon measure and seeing how it looks.

To determine the rotifer count of the aquarium, which should be 25-50 per teaspoon, take a sample

Humbugs such as *Dascyllus aruanus* were among the first fishes bred in home aquaria.

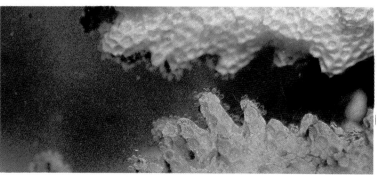

of the water with a dip tube from the middle of the tank, measure out a teaspoonful, and dry it off in a watch glass or similar flat-bottomed glass vessel you can handle easily for counting, then count as before. Do this several times unless the first repetiton gives almost the same result as before. A count of 27 followed by one of 22 would suffice, but one of 31 followed by 12 would not! This all sounds a bit tedious. After a little experience you will be able to estimate densities by eye, but don't depend only on doing that; check yourself now and then by a proper count. Marine fish rearing is so much more demanding than freshwater fish rearing that the relatively rough and ready methods that serve the latter are not good enough. If 5-10 rotifers per drop doesn't sound much in comparison with freshwater infusorial counts that are concerned with animalcules considerably smaller on the average than *B. plicatilis*, consider the fact that a 20-gallon aquarium would contain between 400,000 and 800,000 rotifers. If it had 100 feeding larvae and each ate one rotifer per 5 seconds (a reasonable rate of feeding), they would consume 72,000 per hour, so no wonder there is a need for frequent additions of new food. In fact, as the number of rotifers declines the feeding rate declines also, since the larvae cannot find them so rapidly. If kept at a low feeding rate long enough, the growth-rate of the larvae will also decline.

The larvae should be observed to see that they are feeding adequately. They should be picking away at the rotifers or whatever is being offered, and their tummies should be full and milky colored. As soon as it appears feasible, some newly hatched brine shrimp should be

(Facing page) Spawning in *Dascyllus aruanus* follows the usual course of other damselfishes— spawning display followed by laying of the eggs (in this case attached to coral). The young fish (above) are somewhat different in pattern from the adults but soon develop a full pattern.

tried. It is easy to see if they are being eaten because they will show up pink in the larval stomachs. Once they are accepted, switch to them as a main food but add rotifers for a few days until the smallest larvae have caught up enough to be taking the brine shrimp. Around this time other foods can also be added, although brine shrimp are an adequate diet for the next few weeks. The growing larvae— really fry at a stage somewhere between two and three weeks or even longer, depending on species—can be offered mosquito eggrafts, fine dry foods, and very finely chopped prawn, scallop, etc. Mosquito rafts will hatch on sea water and release the tiny mosquito larvae as a change of diet. If you can gather saltwater mosquito rafts, so much the better; they are to be found floating on high-level tide pools that don't get washed out with each tide and like the freshwater ones, resemble tiny dark spots with a concave surface facing upwards.

As the larvae become fry and start to take on color and a juvenile fish appearance, the problem of raising them is nearly over. They can be transferred to a normal tank as already described and given plenty of growing space and plenty of food, with less worry about water quality and contamination now that they are serviced by a biological filter. If all normal care has been taken, you would now have a tank or tanks of disease-free fishes that have little natural immunity to the various troubles they will very likely encounter if they are transferred to other hands or to the companionship of other fishes, even your own. Be on the alert for trouble once this step is taken, as tank-bred fishes are "tenderer" than wild-caught

ones. They have not been exposed to diseases to anything like the extent of a wild fish and are not the tough survivors out of a very large number that didn't make it. It will likely take generations of tank breeding before marines attain the status of home-bred freshwater fishes, often selected over years for survival and adaptability to aquarium life.

DR. BURGESS'S ATLAS OF MARINE AQUARIUM FISHES
By Dr. Warren E. Burgess,
Dr. Herbert R. Axelrod,
Raymond E. Hunziker III,
H-1100; ISBN 0-86622-896-9
This books contains over 4000 full-color photos, more than any other marine aquarium book ever published. It shows in full color not only the popular aquarium fishes but also the oddballs and weirdos, the large seaquarium type fishes, both warmwater and coldwater species. In short, this book has it all, and it has it in a format that provides maximum utility to readers. The book supplies the most up-to-date scientific names, and the captions indicate the family, range, size, and optimum aquarium conditions as well. Also included are family by family write-ups on the aquarium care of these fishes.
Hard cover, 8½ x 11", 736 pages
Contains over 4000 full-color photos

SALTWATER AQUARIUM FISHES,
New Edition
By Dr. Herbert R. Axelrod and
Dr. Warren E. Burgess
ISBN 0-86622-399-8
T.F.H. H-914
A very complete book, for the medium-level aquarist who has one or two saltwater tanks and wants to know the best fishes to keep and the best techniques for keeping them. This book has gone through a number of revisions and each edition brings many changes. Modern and up-to-date, and very popular with beginning and medium-level marine aquarists. Written on a high school level, the text covers the fishes on a family-by-family basis for the convenience of readers. A full chapter is devoted to the coverage of fascinating marine invertebrates. Contains over 400 full-color photos.
Hardcover, 5½ x 8½, 288 pages
Now revised so that every photo in the book (more than 400) is a full-color photo.

EXOTIC MARINE FISHES
By Dr. Herbert R. Axelrod,
Dr. Warren E. Burgess and Dr.
Cliff W. Emmens
ISBN 0-87666-103-7
TFH H-938L (Looseleaf)
ISBN 0-87666-103-7
TFH H-938 (Hardbound, non-looseleaf)
For the avid marine aquarist who has one or more tanks in his home. Covers setup and maintenance of saltwater tanks, but the main thrust of the book is its catalog of fish species, describing (and showing in full color) hundreds of species. Complete and authoritative; if the aquarist wants only one book on the subject, this is it. High school level.
Hardcover and looseleaf, 5½ x 8½",
608 pages 88 black and white photos,
477 color photos

THE ENCYCLOPEDIA OF MARINE INVERTEBRATES
By a panel of experts, each specializing in individual phyla
ISBN 0-87666-495-8
TFH H-951
This excellent and enormously colorful book ranges widely over the invertebrate field and provides detailed information on the natural history and taxonomy of every invertebrate group of interest to marine aquarists. A superb compilation of vital information and beautiful photos, this book is also an excellent identification guide.
Hard cover, 5½ x 8", 736 pages
Over 600 full-color photos, many line drawings.

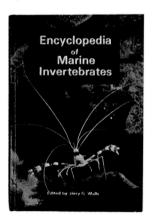

Index

Aeoliscus punctatus, 104
Aeoliscus strigatus, 104
Aeromonas liquefaciens, 101
Algae, 49, 79, 84, 133, 134–136, 145, 155, 182, 184
Amphiprion bicinctus, 165
Amphiprion chrysopterus, 165, 177
Amphiprion clarkii, 165
Amphiprion ephippium, 165
Amphiprion frenatus, 165
Amphiprion mccullochi, 165
Amphiprion melanopus, 165
Amphiprion ocellaris, 75
Amphiprion periderion, 165
Amphiprion rubrocinctus, 165
Amphiprion sandaracinos, 165
Amphiprion thiellei, 165
Amphiprion tricinctus, 165
Amyloodinium ocellatum, 116, 117
Anemonefishes, 6, 48, 57, 70, 74, 75, 76, 86, 92–93, 115, 144, 163, 164, 170, 171, 176, 178, 182
Anemones, 5, 6, 12, 13, 18, 19, 26, 59, 92, 137, 141, 144
Angelfishes, 6, 48, 69, 70, 74, 75, 77, 94, 95, 124, 170, 172
Anglerfishes, 48, 74, 94, 164, 170, 172, 175
Antennarius pictus, 173
Antennarius striatus, 173
Aspidontus taeniatus, 96
Aulonocara nyassae, 22
Australian seahorse, 171
Balistoides conspicillum, 74
Belted Sandfish, 163
Blackspot, 104
Blennies, 6, 68, 75, 78, 93
Blue Devil, 178
Brine shrimp, 70, 79, 81–85, 87, 91, 142, 171, 181, 189
Brittlestars, 144
Brown algae, 146, 149–150
Butterflyfishes, 75, 93, 95, 172
Capoeta hulstaerti, 22

Carcarhinus melanopterus, 10
Cardinalfishes, 77, 170, 179
Catfishes, 45
Caulerpa, 26, 49, 79, 84, 146, 147, 148
Centropyge loriculus, 73
Cephalopholis miniatus, 72
Chaetodons, 6, 48, 75, 76, 78, 93, 95, 96, 101
Chaetodon larvatus, 160
Chaetodon leucopleura, 160
Chaetodon madagascariensis, 160
Chaetodon melanotus, 160
Chaetodon meyeri, 76, 77
Chaetodon ocellicaudus, 160
Chaetodon ornatissimus, 76, 77
Chaetodon paucificiatus, 160
Chaetodon semilarvatus, 102
Chaetodon triangulum, 160
Chondrococcus columnaris, 114
Chondrus, 49
Chromis atripectoralis, 183
Chromis viridis, 178
Chrysiptera evanea, 161
Chrysiptera flavipinnis, 161
Chrysiptera parasema, 164
Chrysiptera rollandi, 164
Chrysiptera starcki, 178
Chrysiptera taupou, 161
Clams, 13, 90
Cleaner blennies, 96
Cleaner gobies, 95
Cleaner wrasses, 73, 75, 95, 96, 128
Clown Sweetlips, 132
Clown Triggers, 74
Coelenterate pox, 14
Coelenterates, 10, 141
Copepods, 5, 101, 182, 185, 186
Coral, 5, 11, 12, 18, 25, 26, 40, 41, 52, 55, 60, 77, 78, 115, 136, 138, 139, 151
Crabs, 6, 11, 13, 16, 25, 86, 88, 143

Crustaceans, 17, 59, 75, 77, 88, 93, 103, 138, 141, 143, 181
Cryptocaryon irritans, 65, 117
Cuttlefishes, 144
Cyanide, 64–65
Cyphoma gibbosa, 156
Damselfishes, 6, 48, 57, 70, 75, 76, 86, 139, 163, 164, 178, 183, 189
Daphnia, 85, 91
Dascyllus, 45, 69
Dascyllus aruanus, 187, 189
Demersal fishes, 179, 180
Demoiselles, 78, 177 Dropsy, 131
Echinoderms, 11, 138
Eels, 45, 74, 116
Electric Rays, 70
Featherstars, 144
File fishes, 75
Fishes in general, 157–161
Fish lice, 126, 127
Flamingo tongue, 156
Flukes, 124–125, 132
Fronded algae, 26
Fungi, 119, 122
Giant clams, 151, 155
Gaterin chaetodonides, 132
Globiosoma evelynae, 159
Globiosoma genie, 159
Globiosoma illecebrosum, 159
Globiosoma randalli, 159
Globiosoma xanthiprora, 159
Glugea, 118
Gobies, 6, 57, 68, 75, 78, 93
Gobiosoma oceanops, 95
Green algae, 18, 25, 26, 27, 146, 147–149
Groupers, 68, 70, 72, 74, 87, 163
Halicorne, 157
Halimeda opuntia, 157
Harlequin Bass, 172
Heniochus, 102
Henneguya, 100
Hermit crabs, 92